PLANTS & GARDENS

BROOKLYN BOTANIC GARDEN RECORD

A NEW LOOK AT VEGETABLES

1993

Plants & Gardens, Brooklyn Botanic Garden Record (ISSN 0362-5850)

is published quarterly at 1000 Washington Ave., Brooklyn, N.Y. 11225, by the **Brooklyn Botanic Garden, Inc.**

Subscription included in Botanic Garden membership dues ($25.00 per year).

ISBN # 0-945352-78-6

Brooklyn Botanic Garden

STAFF FOR THIS EDITION:

ANNE RAVER, GUEST EDITOR

BARBARA B. PESCH, DIRECTOR OF PUBLICATIONS

JANET MARINELLI, EDITOR

AND THE EDITORIAL COMMITTEE OF THE BROOKLYN BOTANIC GARDEN

BEKKA LINDSTROM, ART DIRECTOR

JUDITH D. ZUK, PRESIDENT, BROOKLYN BOTANIC GARDEN

ELIZABETH SCHOLTZ, DIRECTOR EMERITUS, BROOKLYN BOTANIC GARDEN

STEPHEN K-M. TIM, VICE PRESIDENT, SCIENCE & PUBLICATIONS

FRONT AND BACK COVER: PHOTOGRAPHS BY ROSALIND CREASY

JERRY PAVIA

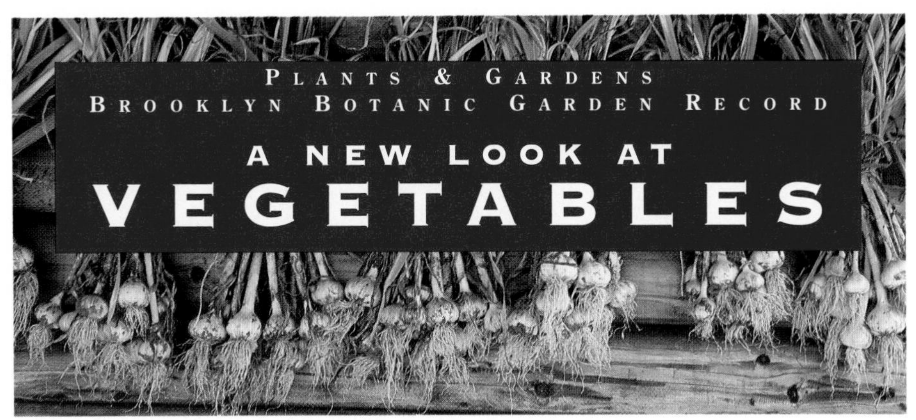

VOL. 49, NO. 1, SPRING 1993

HANDBOOK #134

Introduction	*Anne Raver*	4
The Perfect Vegetable Patch	*Karan Davis Cutler*	8
A Well-Fed Soil	*Eliot Coleman*	14
To Dig or Not to Dig	*Lee Reich*	18
Full Tilth Boogie: Composting Comes of Age	*Phil Tietz*	23
Starting Seeds	*Shepherd Ogden*	29
Common Sense Pest Control	*Sheila Daar*	32
The Predator Patrol: Putting Good Bugs to Work in the Garden	*Cass Peterson*	38
The Bambi Factor	*Walter Chandoha*	42
Drip Irrigation for Vegetables	*Robert Kourik*	48
Great Greens	*Terry Keller*	52
The Upscale Spud	*Rosalind Creasy*	56
The Art and the Science of Tomatoes	*Warren Schultz*	60
Beyond the Green Bell Pepper	*Renee Shepherd*	66
Up With Eggplants	*Alan Gorkin*	72
Essential Herbs	*Tovah Martin*	76
Edible Flowers	*Cathy Wilkinson Barash*	80
A Cornucopia of Corns	*Felder Rushing*	86
The Cultivated Carrot	*Tovah Martin*	90
Index		94

Introduction

ROSALIND CREASY

I used to think that as I became a more sophisticated gardener, I would lose my passion for vegetables. The tomato patch would be given over to cutting-edge perennials. Instead of beans, I'd be growing *Heuchera* 'Palace Purple'. Lilies in place of corn.

Well, it hasn't happened. The farmer in me craves to plow yet another field — for more exotic vegetables. This year I grew grain amaranth and have just harvested its nutty kernels. My Black Aztec corn is drying on the vine — but can I grind it in my Cuisinart?

Every year a new crop of garden catalogs includes new and improved vegetable varieties that are worth considering.

My garlic was attacked by root maggots, but I salvaged enough to want to do it perfectly next year. (A sandier patch perhaps?) A trip to Hatch, New Mexico, made me a chile addict — and I returned with the seeds of half a dozen different varieties, some red, some green, some purple, some yellow and orange. And most of them hot enough to leave the guests calling for beer — and then more chile peppers.

There is no greater pleasure than a plateful of juicy ripe tomatoes marinated in olive oil and fresh basil, unless it's potatoes dug right from the ground or green beans picked when they are pencil-thin, steamed just enough to heighten the flavor, and tossed with dill.

In late summer, I stand in the kitchen barefoot, cooking up sauces, sauteing squashes and eggplants, running out to the garden for one more handful of thyme, or one more hot pepper, to make perfection. Who needs to go to some fancy restaurant like Orso in New York City when you can dip your crusty bread in olive oil and fresh rosemary, right there on the front porch, listening to the katydids?

Tasting the earth and the sun in your mouth is about as sensual as gardening gets — other than sticking your nose in a lilac bush in May. And another nice thing about vegetables: almost anyone can grow something the first time out. Oh sure, the flea beetles may devour the eggplants and the cabbageworms will get the Brussels sprouts (mine did this year, through my own neglect), but, chances are, some tomato vine will threaten to grow over the house

Pay particular attention to the new disease-resistant varieties. There are more each year. PHOTOS BY JOANNE & JERRY PAVIA

and spill tomatoes into the bedroom windows. Even if you only have a few containers on the terrace, growing a salad garden, a pepper bush, a big pot of herbs, a barrel of 'Early Girls' will deepen the pleasures of summer. To say nothing of the vitamins and nutrients flowing into your system — sans pesticides or additives.

This handbook cannot, of course, offer a complete grounding in the complexities of good growing — but it can lead the way into the vegetable patch to begin a lifelong adventure.

I am an organic gardener and I do not use chemical pesticides, or peat moss, because it is a dwindling resource. Many growers still do — especially those who start seedlings commercially and cannot risk a plague of viruses that might course through soil that has not been thoroughly sterilized. But it is up to all gardeners to look for alternatives, to read and question the corporations mining the peat bogs and making the pesticides and the chemical fertilizers, and insisting on their efficacy. Instead of seeing those bugs in your garden as horrible intruders, think of them as your teachers — even as they eat holes in your cabbage leaves. Gardening is, after all, a balancing act with the earth. As good farmers tell their children, give to the earth, and it will give back to you.

ANNE RAVER
Guest Editor

ANNE RAVER *writes about gardening and the environment for* The New York Times.

The Perfect Vegetable Patch

BY KARAN DAVIS CUTLER

WALTER CHANDOHA

Location, location, location — it's the mantra of real estate agents. And of experienced vegetable growers, I learned by struggling in my first garden, a small plot shaded by silver maples. My crops, slow to ripen, were plagued by leaf spot, powdery mildew and a half-dozen other fungal diseases that thrive in dim light.

Knowing where to site a vegetable garden requires only knowing what vegetables like, and that's no mystery. Thomas Hill, in the first garden book written in English (*The Gardener's Labyrinth*, 1577), set forth most of the basics — matters of light, air, water and soil — often quoting the Roman Pliny who, Hill observed, lived when a garden was no more than "a smal & simple inclosure of ground." Gardens may have changed, but those fundamental requirements haven't.

Yet few of us have the sheltered, sunny, well-drained southwest-sloping 20 x 30-foot plot of organically rich sandy loam located near an infinite source of water that the experts, classical and contemporary, recommend. When faced with the facts of horticultural life — moss-covered ledge impersonating soil, ancient oaks blocking all light from noon until night, offspring demanding badminton courts — most would-be gardeners must surrender, a definition of compromise once promoted by Emerson. The news isn't all bad, however: it is possible to farm a flawed location, to compromise *and* produce vegetables.

"Let there be light" was a universal admonition but surely one with the gardener in mind. When selecting a spot to grow vegetables, sunshine is the most critical and least negotiable ingredient. At a minimum, most vegetables require six hours of sun; eight hours is better; all day is best.

KARAN DAVIS CUTLER *is senior editor of* Harrowsmith Country Life. *She gardens between the rocks of Northern Vermont.*

CROP YIELDS

(when spaced per seed-packet directions)

VEGETABLE	AVERAGE YIELD PER 100-FT ROW
Beets	125 pounds
Broccoli	80 pounds
Bush snap beans	120 pounds
Pole snap beans	150 pounds
Cabbage	150 pounds
Carrots	100 pounds
Corn	10 dozen ears
Cucumbers	120 pounds
Leaf lettuce	50 pounds
Onions	85 pounds
Peas	20 pounds
Peppers (green)	50 pounds
Potatoes	100 pounds
Radishes	100 bunches
Spinach	40 pounds
Squash, summer	150 pounds
Squash, winter	100 pounds
Tomatoes	100 pounds

WALTER CHANDOHA

The ideal vegetable patch is a sheltered, sunny, well-drained, southwest-sloping plot of organically rich sandy loam located near an infinite source of water — something few of us have.

There are a few edibles (see "Vegetables for Partial Sun," page 12) that tolerate darkish settings, mostly leaf and root crops, and in relentlessly hot regions afternoon shade is an asset. But sun-poor gardens are slow to warm in spring, susceptible to diseases and notorious for producing spindly plants and harvests. All the aluminum foil, white mulch, mirrors and other tricks used by those saddled with beclouded gardens won't produce even a basket of second-rate beefsteak tomatoes.

The ideal exposure, especially in the North, is a sloping one, a southern or southwestern site. The greater the incline — the more perpendicular it is to the sun's angle — the more quickly the soil heats up and the smaller the danger of frost. A modestly sloped garden in Cleveland, for example, gets nearly the same solar exposure as a level garden in Cincinnati, 250 miles to the south.

Plants need fresh air as well as sun — Hill warned that "evil aire ... doth not only annoy and corrupt the plants ... but choke and dul the spirits of men" — so avoid placing your garden in a low area, where air stagnates, or a spot that is tightly enclosed. Some protection from the wind is desirable, though. A windward barrier — trees, hedge, fence, wall, building or other windbreak — not only protects plants from damage, it reduces soil erosion, moisture and heat loss. Live barriers, be warned, can

WALTER CHANDOHA

It is possible to produce vegetables in a less-than-perfect location. Sunshine is the most critical ingredient. Most vegetables require a minimum of six solid hours of sun per day.

grow to shade gardens, and if shrubs are planted too close — less than six feet — their roots will filch nutrients needed by your corn and cabbages. (Stay beyond the drip lines, or root zone, of trees, especially shallow-rooted species like willows and maples, and far, far away from black walnut trees, *Juglans nigra*, whose roots exude juglone, a chemical that is toxic to tomatoes and other vegetables.)

Ground where "the watriness shall exceed," as Hill put it, is an inhospitable place for a garden. Plant roots need air as well as moisture, and waterlogged soil is air-poor soil. Solving the drainage problem may be only a matter of opening compacted subsoil or improving porosity by adding organic matter, but seriously wet locations likely have to be drained. That can be a whopping undertaking, one involving ditches, gravel-filled trenches, pipes, tiles, dry wells and other expensive options.

Some gardeners can circumvent poor-draining soil by creating raised beds (conversely, gardeners in arid regions create sunken beds to collect much-needed moisture). Better still is to avoid the problem altogether by testing for drainage before choosing a home for your garden. One easy way to measure soil porosity is to dig a hole, 1 x 1 foot, and fill it with water. Refill the hole the next day and keep track of how quickly it drains: longer than ten hours is a signal that you should plant elsewhere.

Soil, while it receives the most attention from the experts, is the least important consideration when siting a garden because soil is the least immutable. Fertility can be improved; texture and structure can be altered; pH can be adjusted. Over the years, goodly applications of horse manure, straw, leaves and grass clippings have transformed my thin Vermont dirt into "fat earth," the country term for fertile soil. Adding organic matter — the horticultural rendering of Lydia Pinkham's Vegetable Compound — is the best way to cure the ills of soil, sand or clay.

What else? Not all wetness is bad, for one. If your garden is unlikely to receive the weekly inch of rain that is the rule-of-thumb charge for vegetables, try to locate it near a source of water. And try not to locate it near a busy road or the foundation of an old building: automobile exhaust and paint are two common sources of lead contamination. Not setting tomatoes across the shortcut to the garage or two feet from home plate seems obvious.

I like having my vegetables near the house, where I am encouraged to tend them and then can admire my work. For some gardeners, this means turning the front lawn into a vegetable patch or creating an "edible landscape," including vegetables and fruits in the ornamental plantings. While not a new idea — people have been tucking edibles among their dooryard flowers for centuries — it is an increasingly popular one, driven by our smaller and smaller yards. Its culmination is container gardening, raising vegetables in pots and boxes, which allows one to be both farmer and landless. Ever sensitive to trends, seed companies now offer a slew of hanging tomatoes and bush melons to compete with fuchsias and petunias for a place on the deck.

For those of us from the old school — and blessed with plenty of space — nothing is as comforting as the traditional vegetable garden, filled with right angles and order. Vegetables ought to grow in rows, just as dogs ought to come when they're called. I run rows east to west to maximize light, but row direction should depend on the lay of the land. In either case, plant tall crops like corn, tomatoes and pole beans

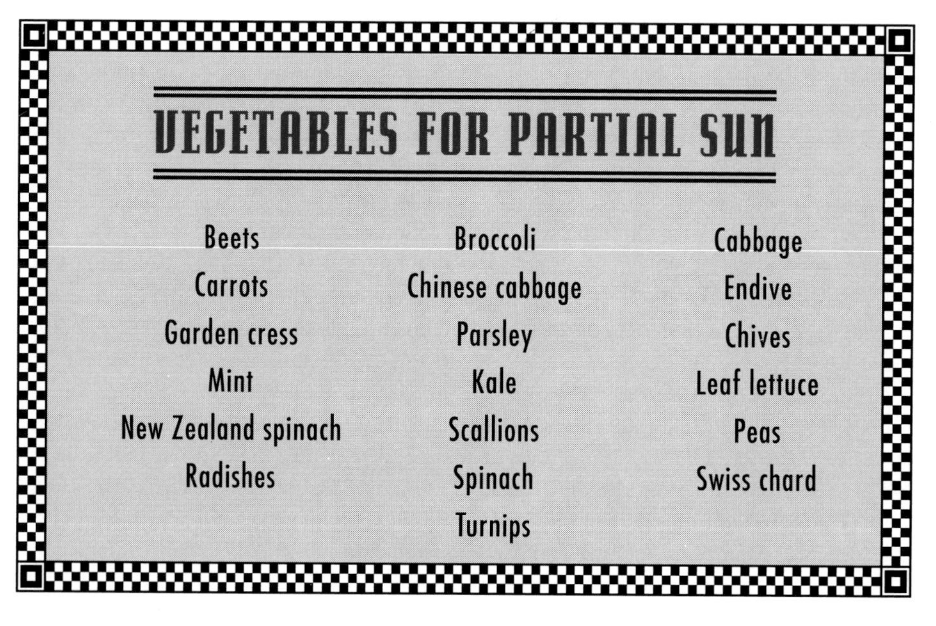

VEGETABLES FOR PARTIAL SUN

Beets	Broccoli	Cabbage
Carrots	Chinese cabbage	Endive
Garden cress	Parsley	Chives
Mint	Kale	Leaf lettuce
New Zealand spinach	Scallions	Peas
Radishes	Spinach	Swiss chard
	Turnips	

WALTER CHANDOHA

Raised beds warm earlier in the spring, enabling you to get your garden off to an early start. They also have good drainage and are easy to tend. Run the rows east to west for maximum sunlight.

on the north side of the garden where they won't shade shorter plants. (If your site slopes significantly, orient the rows across rather than down the incline.)

How large a garden is up to you and what your space permits. A well-planned, well-tended 600 square feet should keep a family of four flooded with vegetables, but even plots as tiny as nine square feet can provide a summerful of salads. There are plenty of space-saving techniques — wide-row, or intensive, planting, intercropping, succession planting, trellising and more — that will increase harvests without requiring a larger garden. The best plan is to plant what you want to eat, keeping in mind that some vegetables are more "space efficient" than others. (See "Crop Yields," page 9) Based on yields per square foot, value per pound and seed-to-harvest time, the National Garden Bureau rated tomatoes as the most space-efficient vegetable, followed by scallions, leaf lettuce, summer squash, snow peas, onions, beans and beets. Sweet corn, melons and pumpkins, alas, were at the bottom of the list.

Once the site is chosen, the real fun begins. Tilling, sowing, weeding. Discovering the subtleties of place and plant. Gourds "planted in the ashes of men's bones, and watered with oile, yeeldeth fruit by the ninth day," Thomas Hill declared. Whether the rows ran east to west is unrecorded.

A Well-Fed Soil

BY ELIOT COLEMAN

The best model for creating fertile garden soil is the natural world. That's as it should be. The home garden is only a human-managed microcosm of the larger natural garden that surrounds it. Whether forest or prairie or highland or lowland, soil fertility in the natural world is maintained and renewed by the recycling of plant and animal residues. This recycling is a biological process, which means that the most important contributors to natural soil fertility are alive and they are not us. They are the population of living creatures in the soil.

The numbers of soil creatures are impressive. Research has shown that an area of fertile soil supports as great a weight of organisms "grazing" under the surface as it does feeding on the plant growth above. That means if one 2000-pound steer can find a year's worth of food from an acre of grass-legume pasture, there will be an additional ton of valuable

---

ELIOT COLEMAN'S *latest book,* Four Season Harvest *(Chelsea Green Publishing Company, 1992), celebrates the potential of eating year-round fresh food from your home garden no matter where you and your soil livestock live.*

"livestock" beneath the ground. Because most of these creatures are microscopic, we are talking about billions and billions of soil inhabitants.

The best approach to creating a fertile soil is to think of yourself as a farmer for *underground* livestock. They need to be fed, housed and cared for. These livestock are much more varied than the steer. They range in size from the microscopic and unfamiliar (bacteria) to the visible and familiar (earthworms). In between is an enormous population of micro- and not-so-micro-organisms whose activities and life processes are as fascinating a tale of ecological design and balance as can be told. I once watched a specialist on soil creatures perform the seemingly impossible task of holding the rapt attention of a roomful of teenagers by showing slides and telling tales of the endlessly interrelated and meticulously choreographed activities of these creatures. The students were entranced because the subject matter was like a trip to another planet. They were peeking into the secret world of nature.

The idea of caring for soil livestock also helps illuminate the difference between a natural and a synthetic approach to fertilizing the garden. In the synthetic approach fertilizers are compounded chemically to put a limited number of nutrients in a solu-

CHRISTOPHER GIAM

The best approach to creating fertility is to think of yourself as a farmer for *underground* livestock, the billions of creatures living in the soil.

ble form within reach of plant roots. The idea is to take charge of the garden and start "feeding the plants" directly with preprocessed food. In the natural approach you the gardener work with natural soil processes to make the plant-food potential of the soil available to plants. You do that by adding organic matter and natural rock minerals to the soil. This is usually called "feeding the soil" as opposed to "feeding the plants." But what you are really doing is feeding the soil livestock and that's why it works. As they "do what comes naturally," they break down organic matter, converting nutrients into forms plants need.

Growing a garden by depending on the natural processes of the soil organisms also makes allowance for our ignorance. No matter how much we may think we know about plant nutrition there will always be subtle factors in the natural system that we haven't figured out and that are important for plant health. The soil organisms have much more experience (millions of years worth) than we humans do in providing all the necessary nutrients. Furthermore, plants have evolved to thrive on nutrients in just the forms that the soil livestock provide them.

The environment of the soil world in which these organisms live is obviously important to their survival. Soil is composed of four parts — mineral particles, air, water and organic matter. The mineral particles provide both structure and nutrition. They range in size from the smallest, clay, to silt, to the largest, sand. Different size particles result in different types of soil. The nutrition provided by the mineral particles depends on the parent rock from which they originated. Some soils have a relatively ideal balance of the minerals on which most vegetable plants thrive. Others, like my soil in coastal Maine, which is low in calcium

The many organisms that comprise the soil livestock break down organic matter, converting nutrients into the forms that plants need.

and phosphorus, will benefit from additions of such minerals as phosphate rock and limestone. Because the soil livestock necessary for good vegetable growth flourish at a pH of around 6.5 to 6.8, the first step in caring for them is to counteract soil acidity.

Much of a fertile soil is composed not of what you can see but what you can see *through*. A good air supply is crucial for the health and survival of both the soil organisms and the roots of garden plants. A compacted, airless soil (like the path where the kids cut across the corner of the lawn) has a greatly lowered capacity for plant growth. An overly wet soil will also kill off the valuable soil livestock through lack of air. Some provision for drainage is obviously necessary. An overly dry soil, however, is not a congenial habitat, either. The moisture in the soil is most favorable to soil organisms when it is in balance — not too much and not too little.

CHRISTOPHER GIAM

But it is the organic matter that is the consistent thread linking soil materials and soil inhabitants. Like the leaf mold on the forest floor and the loam on the prairie it is the key to soil fertility and the well-being of the soil livestock. Organic matter opens up heavy clay soils and gives structure to light sands, making them both better habitats for the soil livestock. The decomposition products of organic matter glue together the natural crumb structure of a fertile soil, so vital in allowing air to enter. The water-holding capacity of organic matter assures ample moisture. Its porosity assures good drainage. And finally, organic matter is the food for all those underground creatures. The numerous byproducts of the consumption and decomposition of that organic matter are the nutrients needed for good plant growth. No life in the soil, no life *on* the soil.

The best way to add organic matter is to spread it on the soil surface in the fall, just as nature does. Mix it in shallowly if you wish. If the organic wastes have been partially predigested in a compost heap, so much the better, but even raw residues will be beneficial. The covering of organic matter protects the soil and your soil livestock against the cold of winter. The soil livestock will begin to transform the organic matter right away. In spring any undecomposed residues can either be raked off and returned to the compost heap or left as a mulch, depending on the seeds to be sown, the plants to be set out or your preferred style of gardening.

So take your inspiration from nature and recycle as much organic residue (weeds, clippings, straw, vegetable waste, manure, leaves) as possible in your garden. Go out there every fall and feed your livestock. Well-fed soil critters mean well-fed vegetable plants. "Ok, gang, it's feeding time...eat up...more where that came from...Anything else I can do for you?...All set?...Bon appetit!"

To Dig or Not to Dig

BY LEE REICH

Spring soil preparation separates gardeners into two camps: the diggers and the non-diggers. Over the years, the ranks of non-diggers have grown — and for some good reasons. Nonetheless, some gardeners still insist on partaking of the annual ritual of turning over the soil. Before discussing the virtues of not digging, let's make short work of the digging camp.

What is the rationale for turning over the soil each year with a Rototiller, spade or garden fork? Digging aerates the soil, and is the way to mix in quantities of organic materials such as leaves, sawdust or manure. There also is a psychological benefit to digging the garden each spring. The hard work stirs the blood as well as the soil. And a freshly prepared seed bed is like a clean slate, with last year's mistakes erased.

LEE REICH *has a PhD in horticulture and a masters degree in soil science. He is the author of* Uncommon Fruits Worthy of Attention *(1991) and* A Northeast Gardener's Year *(1992), both published by Addison-Wesley Publishing Co., and is also a garden consultant. He lives in New Paltz, New York.*

If you insist on digging, remember these two basic rules: don't dig too soon and don't dig too much.

The not-too-soon rule: wait for the soil to dry out a little before tilling in spring. Working a wet soil, especially one that is clayey, ruins its crystalline structure. The soil might be good for sculpture, but it's poor for vegetable plants. On the other hand, do not attempt to work the soil when it is bone-dry. Tilling such a soil leaves large, rock-hard clods.

The way to determine if a soil is ready to be tilled is to squeeze a handful of it. If it crumbles easily, it is ready to till; if it wads up, it needs to dry out more. If the soil contains just the right amount of water, a shovel or spading fork will slide in easily. As each shovel or forkful is lifted and turned, the clods will start to break apart from their own weight. Following spadework, gentle coaxing with a garden rake easily crumbles larger clods into smaller aggregates for a seed bed.

Although offering less psychological satisfaction (because the urge to till is less in the autumn), autumn tillage has some advantages over spring tillage. In autumn, the ground is usually moist, but not sodden. And tilling in autumn leaves one less task to do amidst the flurry of spring gardening activities. If you do till in autumn, leave the soil rough — pounding from autumn and early spring rains will wash away and ruin the surface of a smooth, finely tilled soil. Let the action of freezing and thawing during the cold months do your work, breaking up large clods. Come spring, you will only need to rake the soil lightly, just before planting, to prepare the seed bed.

The not-too-much rule: if you do dig, do not reduce the soil to a fine powder. Soil particles bind together into aggregates, called peds. Plant roots need both water and air to function, and these peds give the soil a variety of pore sizes. The smaller pores retain water due to capillary attraction, but the larger pores cannot hold capillary water, so quickly drain water and fill with air. Especially with a power Rototiller, it is too easy to pulverize the peds. Control the urge to run the tiller up and down the rows until the soil is like dust. Nothing beats tillage with a shovel, followed by raking, for exercise and for tempering the tendency to overwork the soil.

But is all this work necessary, or even beneficial, for plants? Turning the soil is an age-old tradition that dies hard. Edward Faulkner laid the first serious challenge to this annual ritual in his book, *Plowman's Folly* (1944). Ruth Stout further popularized the concept of no-digging in her book, *How to Have a Green Thumb Without an Aching Back* (1955).

The facts speak for themselves: churning the soil charges it with oxygen, a process that rapidly burns up valuable organic matter. Churning the soil also destroys large channels left by earthworms and old roots, as well as small capillary connections throughout the soil. This connecting system of large and small pores is what moves air and water throughout the soil — up, down and sideways. If you want proof, take a look at the lush plant growth along roadsides and in pastures, where the soil is undisturbed; even midsummer's heat hardly causes these plants to droop. Digging also brings to the surface dormant weed seeds, just waiting for a bit of light and perhaps a bit more air to infuse them with life. And finally, gardeners who forego digging need not delay spring planting until the soil dries.

So what do you do if you do not dig? The best reason to till the soil is to aerate it, but if you avoid compacting the soil, you avoid the need to dig it. To avoid compacting the soil, set up the garden in permanent beds, and never walk on them. Beds three to four feet wide are narrow enough to plant, weed and harvest from paths bordering the beds. (These beds need not be raised beds, which are useful only where drainage is poor, but otherwise tend to dry out too readily in summer.) Eighteen inches is wide enough for the paths, which you can easily keep weed-free with a covering of any of a num-

LEE REICH

In the no-dig garden it is important to avoid compacting the soil. Set up the garden in permanent beds and never walk through them.

ber of materials such as wood chips, sawdust, leaves or even a few layers of newspaper topped by chips, sawdust or leaves.

Starting a no-dig garden is simple and quicker than beginning by digging up the ground. For a site initially in lawn or weeds, the conventional approach is to turn over the soil, wait a couple of weeks for the vegetation to decompose, then turn the soil again and rake it smooth. For the no-dig approach, begin by mowing the proposed garden area, laying out the beds, then sprinkling some fertilizer and, if needed, lime over them. Next, cover the growing beds with four layers of newspaper. This smothers existing vegetation, which, along with the newspaper itself, eventually decomposes to form humus. On top of the newspaper, apply a few inches of compost. Presto! You're ready to plant.

Compost is the ideal surface covering, but other organic materials such as leaves, manure and straw will suffice if you are going to set transplants rather than plant seeds the first season. Make holes through the newspaper, if necessary, to set large transplants in the ground.

Even in a no-dig garden, there is one situation that calls for digging — a site where the soil is a sticky clay that needs quick improvement. But even in this case, digging is needed only once — in the initial preparation of the ground. The goal here is to mix a massive quantity of organic matter into the soil, preferably material that has a lasting effect because it resists decay. Both peat moss and sawdust are resistant to decay, though from an environmental point of view the latter, a "waste" product waiting to be recycled, is preferable to the former, which must be mined from bogs. (In subsequent years, surface additions of organic matter will work their way down into the soil with the help of earthworms.)

LEE REICH

The author's no-dig garden, shown in spring on the opposite page, is still showing healthy, lush growth in September, above.

LEE REICH

In late fall the mulched pathways are visible. Keep the beds three to four feet wide for easy planting and upkeep. Paths should be 18 inches wide.

Wait until the soil is just moist and spread nine cubic feet of peat moss or twelve bushels of sawdust over every 100 square feet of bed. To counteract the acidity of either of these materials, also spread 20 pounds of ground limestone over the area. Soil microorganisms that slowly decompose sawdust need extra nitrogen for the job, so add 40 pounds of a fertilizer having about 10 percent nitrogen (such as soybean meal or 10-10-10) to the above quantity of sawdust. Thoroughly mix the peat or sawdust, and additional materials, into the top six to 12 inches of soil with a tiller, garden spade or fork, and you are ready to plant. Then recycle these tools by passing them along to a still-digging gardening friend — *you* won't be needing them anymore.

Maintain no-dig beds each year by applying fertilizer and lime, if needed, then laying down a one- to two-inch blanket of weed-free compost or other organic material. The weed-free blanket adds additional fertility and smothers most small weeds. Leaving organic matter on the surface of the soil puts it where it does the soil and plants the most good. This layer protects the surface from washing away in pelting rain and insulates lower layers from searing sun. Over time, earthworms eventually "till" some of this organic matter more deeply into the soil.

Even a no-dig garden needs some weeding, especially for the first two years. Dig out large weeds individually with a trowel, roots and all, being careful not to churn the soil excessively. Cut the tops off any small weeds that do grow by skimming the soil surface with a sharp, hand-held hoe. After a couple of years of regular weeding and not turning the soil (hence, not "sowing" weed seeds), you'll notice that the weeds are sparse enough that weeding becomes a pleasant diversion, not an incessant chore.

LEE REICH

A cross section of a no-dig bed shows soil well aerated with earthworm channels. Every year, apply a one- to two-inch blanket of compost.

Full Tilth Boogie

Composting Comes of Age

BY PHIL TIETZ

With today's shrinking landfills, sanitation budget cuts and struggling recycling programs, the importance of composting looms ever larger. As gardeners we have the opportunity to recycle organic solid waste from start to finish — and utilize the finished product, right in our own communities and our own backyards.

There are a variety of composting techniques suitable for every lifestyle, backyard and budget.

Three-Bin Method

This popular method for backyard composting has cachet. It enables you to boast, "I have a three-bin system," to which other compost-makers will nod approvingly and say, "Yes, yes, very good." Three bins contain the piles of material to be composted; the bins must be at least 3' x 3' x 3' in capacity — one cubic yard. To ease loading and unloading, leave the front of the bins open — then you can simply drive a wheelbarrow right into the bins and dump away. If your compost begins to pile up, just put a plank on the pile and use it as a ramp.

I use a set of three bins that are four feet tall, three feet wide and four feet from front to back. To turn the pile, I rake the compost out onto the ground in front of the bin with a long-handled cultivator, and pitchfork it back into the bin — a lot less

PHIL TIETZ

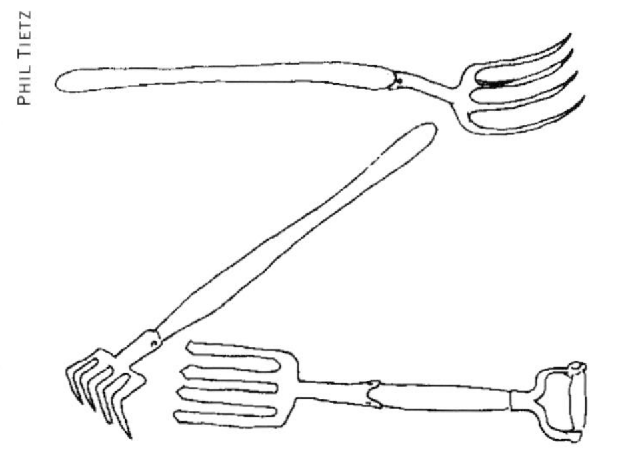

Composting tools: manure fork, cultivator and digging fork.

PHIL TIETZ *started gardening in Kansas City, where he grew roses as a small child and later developed an organic garden. He has been community gardening in New York City since 1978, and is now the assistant director of the Green Guerillas. He operates three different composting sites in the city.*

JUDY WHITE

For loading and turning compost, a long-handled manure fork or short-handled digging fork come in handy.

work than forking it up and over from one bin into another as conventional wisdom dictates; besides, I can keep all three bins working instead of leaving one empty to turn compost over into.

For loading and turning compost a long-handled manure fork is good for extra leverage. It is especially helpful for turning light fibrous material like hay and weeds. Short-handled digging forks with flat tines are better for finely ground materials. It's good to have both pitchforks on hand. Don't get caught standing on the hay you're trying to fork and straining to lift yourself, as Robert Frost described it.

When you fill up your bins with a good mix of materials — carbonaceous (sawdust, woody prunings) and nitrogenous (manure, kitchen scraps, green prunings) — the pile will begin to heat up. This is a sign of intense activity by bacteria and other microorganisms in the pile. As long as you can keep the pile running hot, the compost will break down quickly. Even after the pile cools down, decomposition will continue as long as the pile is turned and watered. Compost manuals advise turning the pile when the temperature tops out around 140 degees F. A compost thermometer with a long probe can be used to monitor the temperature. But unless you're the type who loves gardening gizmos, simply plunge your hand deep into the pile. If it's so hot you have to pull it out quickly, the temperature is topping out! Hot-running piles speed up composting, discourage flies and rodents and kill weed seeds in the mix. Other than these piddling points they're useful only as a matter of pride. Any pile of organic matter, left to its own devices, will eventually rot. So if you don't want to bother with the turning and the rest, and you have the room, just relax and let nature do the work. Indeed, you don't even need the bins. The easiest way to compost is with a free-standing pile, or a large cylinder or square made from wire or snow fencing to keep your pile under control.

Competitive composters take note: to keep the pile running hot you must turn it at least once a week and keep it damp but not soggy. Leave the top of the bins uncovered to let in rain; if it doesn't rain, hand water. Turn the piles every two or three days to speed up the process. For continuous hot-running piles, you may need more nitrogen-rich materials than the average household generates. Green vegetable scraps are available in quantity at greengrocers or supermarkets; if you're environmentally correct and don't have a lawn, you can get grass clippings from family and friends; and you can get horse manure from many local stables.

You can also avoid the turning by simply heaping up compostables and, once the pile cools down, adding redworms for more complete decomposition. They're sold at fishing-bait stores as "red wigglers" and are available through the mail. You can also get them from gardeners who compost. The worms flee piles that are running hot, so be sure to leave small openings between the bins as escape routes.

An alternative technique is to turn the pile once or twice, then grow "green manure" on top. Here's how it's done: after the first few turnings the pile will no longer heat up, as the nitrogen level has gone down. To finish off the pile, cover it with a thin layer of soil or finished compost and plant seeds on top. Plant them really thick! The best seeds to use are rye, wheat, oats and dried beans. You can get these cheap from the local health food store, food co-op, agricultural supply or grocery that sells them in bulk. Grow the greens on top of the pile for about 2 to 3 weeks, letting the redworms do their work in the meantime. When the greens are about 6 inches high,

JUDY WHITE

A rustic log bin adds style to the vegetable plot, but the easiest way to compost is with a free-standing pile.

PHIL TIETZ

The basic three-bin composter: built for speed, not looks.

JUDY WHITE

You can make compost bins out of scrap lumber. Just be sure to "paint" them from time to time with a vegetable-based oil to help prevent rot.

turn the pile as usual. It will be finished after one or two turnings. This technique is great for gardeners without access to a local stable or other source of nitrogen-rich materials.

My bins are made of scrap lumber, with holes drilled in all of the sides for ventilation and worm access. I use two layers of planks on each wall for stability and (some) insulation. The supports are 3" x 3" lumber posts, sunk two feet into the ground with the surrounding soil tamped down. To prevent (or at least delay) rotting of the wood, I paint it with a vegetable-based oil from time to time.

Over the winter, I leave at least one bin filled with organic material. My resident redworms, which are not very hardy in cold weather, winter over in this pile. I continue to run hot compost in the other two bins, but composting slows down during the winter unless there's a constant supply of manure, in addition to covers on the piles to hold in heat. A covering of black plastic will collect solar heat to help warm the piles. In winter the piles should never get soaking wet or they may freeze solid.

Closed-Container Composting with Redworms

Worm composting is a popular indoor method, but you can do it outdoors on a larger scale and with less fuss. A good rodent-proof technique, suitable for people with small backyards: use any metal, wood or plastic container (an ordinary trash can is fine) with holes drilled on the sides and

JUDY WHITE

Many different types of commercial composters are available. The one at left in the photo above, which rotates on its long axis, is designed to to make turning easier.

bottom for ventilation, and elevate the bin on bricks or planks. Because you don't want the pile to heat up too high for the worms, use a container less than one cubic yard in capacity. If the compost still heats up, you'll need to take out some of the mix and put it in a separate container. Then add sawdust or shredded paper to both batches.

Use a mix of sawdust, shredded paper and leaves (airy, absorbent materials) and vegetable scraps and green garden waste (wet materials rich in nutrients). With worm composting, it is essential to chop up the ingredients into small pieces or run them through a blender. Make sure there's enough light, absorbent material so the worms can get plenty of air. The bin needs to be kept moist but not soggy.

PHIL TIETZ

Redworms can be used to compost organic materials indoors or outside. They're available at fishing-bait stores.

When the worm compost is finished (dark and crumbly), dump it out and put some of the worms back in the bin with fresh materials. The remaining worms and compost can be added to the garden.

Again, you'll have to give worms shelter in the winter, either in an indoor worm bin or by "farming them out" to someone with large bins until spring.

Pit Composting

In winter (or when putting in a new garden bed), you can simplify your composting of kitchen waste by simply burying it in the garden. You need to dig it in at least two feet deep (remember that we now have rabid raccoons even in Manhattan), and it helps to chop up the material. Putting a thick layer of mulch on top of the soil will encourage earthworms to help break down the waste, as will adding manure when digging it in. The process takes about four months. Be sure you don't dig into the compost pit before that time, or you'll uncover a smelly mess!

Commercial Composting Units

Many different compost containers are available commercially. There are tumblers made as drums or barrels that rotate either horizontally or end over end. These are hard to load and unload; you may need a brace to hold the bin in position while you fill it. Another popular model is a cube-shaped bin with vents sold under brand names like "Soilsaver" or "Biostacker." It's very hard to turn the compost inside these units; you have to lean over and fork it from above or dismantle the unit and move it over (and this takes extra space). There is also a green-cone model designed for composting slow, small batches; this model depends on use of a bioactivator product and may turn anaerobic and smell bad. The advantage of all these units is that they work where larger piles or a three-bin set won't fit.

COMPOSTING RESOURSES

You can refer to the extremely readable ***Let it Rot!*** by Stu Campbell for more information. This book, along with ***Home Made*** and its excellent compost-bin plans, is available from Storey Communications; call (800) 827-8673 to order books or a catalog. Extensive information can be gleaned from ***The Rodale Book of Composting*** or the super-hefty ***Encyclopedia of Organic Gardening*** from Rodale Press, 33 East Minor Street, Emmaus PA 18098. Neccessary Trading Company carries composting accessories and publishes a composting "bio-bulletin"; call them at (703) 864-5103.

Starting Seeds

BY SHEPHERD OGDEN

When the seed racks go up at the garden center, you know spring can't be far away. Out in the greenhouse, the first bedding plants and vegetables are being sown to sell to gardeners when the frost is finally out of the ground. But garden centers rarely offer a comprehensive selection of plants, particularly vegetables, so if you want something special, you're going to have to start it yourself.

My grandfather started most of his transplants directly in the rich soil of a cinderblock cold frame set deep into the south side of a soil berm. When they reached the proper size and the weather was right, he would stop what he was doing and transplant them. Today we are rarely so free with our time, and more likely to start our plants in containers.

Growing transplants in an open tray is one of the most space-efficient techniques, but still requires a lot of time and attention.

SHEPHERD OGDEN *is founder and president of The Cook's Garden, a mail order seed and supply house in Londonderry, Vermont.*

One ingenious alternative is soil blocks, which combine the best of growing in open flats and growing in individual containers because the roots have plenty of room to roam but don't intertwine. Soil blocks are made with a small, hand-held block press directly from a potting mix relatively high in peat. Preformed blocks of peat can also be used, or peat pots which are filled with potting soil and then placed in trays.

Clay pots allow the soil to breathe, but are breakable and expensive; plastic pots cost much less. Really thrifty gardeners start their seedlings in recycled paper cups, tin cans and milk or egg cartons. You can also make pot-rings from strips of scrap paper and cardboard held together with staples or tape. Stand the rings in trays and fill with soil; at planting time, slip off the ring and set the plant. Many professionals use special trays with tapered growing cells for the seedlings — similar to garden six-paks, but reusable.

What to Grow and How

Some vegetables won't survive transplanting and so there's no point in starting the seedlings indoors. The largest group of these is root crops; others, like dill, fennel and Chinese cabbage, will likely bolt if transplanted. Still others such as spinach grow so fast there isn't much point in starting ahead.

But many plants thrive on transplanting. Make it easy on yourself: buy the common things and raise only the kinds you can't buy.

Slow growers like perennial herbs, parsley, celery and celeriac should be started 12 weeks before their intended transplant date; if you have limited space, buy plants. Tomatoes, peppers and eggplants are a little quicker and can be sown eight to ten weeks before the frost-free date; just be sure to keep the flats at 80 degrees F or more until germination. Most annual herbs also require eight to ten weeks. Brassicas, squashes, cucumbers and lettuce need only be started four to six weeks before the last frost date.

To calculate the proper time to start your transplants, count back from the last-frost date for your garden the required number of weeks. Remember that windowsill gardeners should allow the maximum time noted; gardeners with a greenhouse can afford to be a bit late starting, as the plants grow faster. It is better to have small plants at transplant time than large plants.

A Warm, Moist Place

Peat moss is an important component of most potting soils as its fibrous structure holds water and helps bind together the ingredients of the mix until roots fill the container. Unfortunately, peat must be removed from Canadian bogs, and pure compost is rarely of loose enough texture to be sufficiently aerated, so until a good substitute for peat is found, I recommend buying a commercial soilless mix that contains a large proportion of peat. When filling the container, dump the potting mix loosely into the tray until it overflows, then scrape off the excess. Don't pack it down as both seeds and young plant roots need air; it will pack down when you water.

As a general rule, plant seeds three times as deep as they are across. They contain a tiny replica of the parent plant and a supply of food, dried for storage, and readily sprout as soon as conditions permit. The first stage of germination is swelling, and as the seed soaks up water, enzymes trigger digestion of the seed's food stores. The second stage begins after a few hours or a few days (indoors) with the emergence of the seed root, or radicle. Water intake and digestion reach a steady level. The third stage is when the seedling really starts to grow. This changeover takes place quickly and minute feeder roots soon begin to spread through the soil.

Different vegetables need different temperatures to germinate, but warm room temperature (70 to 75 degrees F) is fine for most. Heat lovers can be put on top of the refrigerator, in the wash room or furnace room, or any high shelf in a heated house; cool weather crops can go in the pantry or some cooler room.

Growing Places

The relationship between heat and light is all-important to growing healthy, vigorous vegetable seedlings. As soon as the first shoots show in a flat it should be moved to the brightest place you've got because too much heat without enough light leads to leggy, limp plants. This can happen even on a bright windowsill because the glass filters out part of the spectrum of light, but lets in the orange-red portion which stimulates stem growth.

Watch out especially for a combination of cool temperatures, low light and overwatering: damping-off fungus can attack and level a whole flat of plants before you know

CHRISTOPHER GIAM

it. If some plants wither at the soil line and collapse, remove them immediatley and give the remaining seedlings in the flat bright light and good air circulation — you may be able to rescue the survivors.

Water and fertilizer also need to be kept in proper proportion. Don't fertilize by the calendar, but rather with every third or fourth watering. That way the plants get fertilized in relation to their rate of growth. We use a combination of fish emulsion and liquid seaweed for our seedlings.

Most discoloration of the leaves is a sign of nutrient deficiency: if the whole plant is pale, there is likely a nitrogen shortage; purplish undersides indicate a shortage of phosphorus; bronzing of the edges means not enough potassium. Fish emulsion contains all three elements, so simply raise or lower the amount of fertilizer in response. There are many other elements, called micronutrients, required for proper growth, and liquid seaweed has a high content of these trace minerals.

To acclimate your seedlings to the stresses and strains that await them out in the garden, put them outdoors for increasingly long periods each day to help them adjust. The most obvious stress is temperature — extreme cold or heat can hurt young plants used to being pampered. Wind is just as hard on the plants and can lead to wilting. Less obvious is the need to adjust your plants to unfiltered sun. Seedlings grown in low light have extra chlorophyll in their leaves, and if they are moved right out into the sun, their light-gathering capability is so great that the leaves may overload, then fade and fall.

Remember: good gardeners mimic nature rather than try to overrule her. "The plants grow themselves," an accomplished gardening friend reminded me a few years ago. "We're just here to help."

Common Sense Pest Control

BY SHEILA DAAR

JERRY PAVIA

Today, gardeners are increasingly on the lookout for ways to manage pest problems with little or no use of toxic materials. More and more are relying on a process called "Integrated Pest Management," or "IPM." Principle number one of this approach is that total elimination of garden pests is virtually impossible and rarely desirable. If, for example, all aphids in an area were wiped out, then their natural enemies would also die — leaving a biological vacuum open to invasion by immigrating aphids whose populations would explode in the absence of natural enemies. This syndrome occurs commonly when pesticides are overused.

Instead of eliminating pests, the goal of IPM is to determine at what level pests can be present in your garden without causing unacceptable damage. Armed with this knowledge, you can concentrate on tending to the horticultural needs of your plants and encouraging the presence of natural enemies of pests. Only when these activities don't keep pest numbers within tolerable levels do you resort to more direct pest controls.

Monitoring

When you use an IPM approach you must not only be able to identify pests accurately, you must also learn about the life cycles and behavior of the insects, pathogens, weeds or other pests wreaking havoc in your garden. This process, called monitoring, is key to IPM. There are many things to look for:

Condition of your plants. When stressed from too much or too little water, fertilizer, sun, shade, insufficient drainage and so on, plants are more susceptible to attack by pests. By learning to recognize early signs of stress, you can often correct the horticultural causes and thereby increase your plants' resistance to pest attack.

Evidence of damage and abundance of pests. The mere presence of a pest does not necessarily indicate a problem. It is important to ask yourself: is there any damage? What kind? Holes in the cabbage leaves, for instance? Where on the plant is the damage found? How many pests are present: "none," "a few," "many" or "extremely high numbers"?

Presence of natural enemies. Do you know what feeds on the pest or competes with it? Can you tell how to recognize natural enemies such as lady beetles, syrphid flies or parasitic wasps, and know when they have killed pests? For example, aphids killed by tiny parasitic wasps turn a bronze color, while parasitized immature whiteflies turn black.

Weather and microclimate. How hot, cold, wet, dry or windy is it? Are there specific local conditions such as poor air circulation where the pest problem occurs?

Gardening habits. Scrutinize your regular maintenance activities, because they may be encouraging the pest or discouraging natural enemies: for example, how are your plants watered, fertilized, pruned or mowed? Too much soluble fertilizer encourages aphids and other pests. Watering late in the day can encourage diseases.

SHEILA DAAR *is Executive Director of the Bio-Integral Resource Center (BIRC), and co-author of* Common Sense Pest Control *(Taunton Press, 1991), a 715-page compendium of least-toxic solutions to house and garden pests. For more on IPM techniques and sources of least-toxic products, request a publications catalogue from BIRC, P.O. Box 7414, Berkeley, CA 94707.*

Left: Attract beneficial insects to your garden by growing fennel and other plants of the parsley family.

Wood chips, hay and other organic mulches not only help keep down weeds but also conserve water and enrich the soil as they break down. Mulch can harbor unwanted pests such as slugs, so keep checking for them.

A Hav-a-Hart trap can be used to capture pesky creatures without inflicting injury and move them to a location away from your vegetable plot.

Miscellaneous factors. Some out-of-the-ordinary event may have affected pest levels. If your neighbor backed out of his garage and scraped some bark off your fruit tree several months ago, for instance, it may explain why wood-boring insects are attacking it now.

By staying in touch with these workings of your garden's ecosystem, you will learn to anticipate conditions that can lead to pest problems, enabling you to prevent pest outbreaks or nip them in the bud — before they become serious outbreaks.

Initially, you should monitor on a weekly or biweekly basis during the growing season. Once you've become familiar with the identity, life cycle and behavior of common pests in your garden, you can increase or decrease monitoring frequency depending on the season, pest levels and other factors. The most important thing is to make *regular* inspections of your plants so you can keep track of what is going on. This includes checking plants at different times, since garden organisms operate at various hours of the day and night. Jot down your observations in a garden notebook or pages from a calendar. These written records will enable you to spot pest trends over time, and to use the information to prevent future problems.

A number of tools will aid your monitoring activities. For example, sticky traps in various insect-attracting colors, or traps containing chemical attractants called pheromones, can give you an early warning of pest presence. A magnifying glass or ten-power hand lens can help you see tiny insects and mites, as well as the fruiting bodies of fungi that damage plants. A maximum/minimum thermometer is another useful monitoring tool, as garden pests are greatly influenced by the weather. These tools are widely available at nurseries and hardware stores or from garden catalogs.

Establishing Pest Tolerance Levels

The data you collect from your monitoring activities will enable you to establish tolerance levels for particular pests. This is a three-step process. First, you decide how much aesthetic or economic damage you're willing to tolerate on your plants (for example, up to ten percent damaged leaves). Second, you observe how large the pest population can grow before your tolerance level is exceeded (you find that more than 20 aphids per leaf badly distorts bean plants). And finally, you establish a treatment level that keeps the pest population small enough so it does not cause an unacceptable amount of damage (if aphids have reached an average of ten to 12 per leaf and natural enemies aren't evident, you may need to apply a treatment to prevent them from reaching damaging levels of 20 or more).

Least-Toxic Treatments

When pest numbers threaten to exceed your tolerance level, it's time to intervene to reduce pests to acceptable levels. The most effective approach is to use a combination of techniques to attack the pest at several vulnerable points. Always begin with cultural, physical, mechanical, biological and educational methods to solve pest problems; *chemical* controls, even relatively nontoxic ones, are a last resort and should be used only when non-chemical approaches have proven insufficient to solve the problem.

The following are major IPM strategies and tactics, together with a few real-life examples. They are discussed here in the order in which they should be considered:

Design pests out of the garden. Sawdust applied to vegetable garden paths prevents weed growth by tying up soil nitrogen.

Modify the pest habitat. Fruit trees planted on a slightly elevated soil mound are protected from contact with standing water, which enables pathogens to attack the root crown.

Horticultural controls. Substituting slow-release forms of nitrogen (compost, sewage sludge and so on) for fast-release fertilizer prevents the excessive nitrogen levels in plants which attract aphids, scales, thrips and other pests.

Physical and mechanical controls. Hand-held vaccuums can suck up whiteflies; fabric row covers can exclude insects and birds from rows of vegetables; copper strips can prevent slugs and snails from reaching plants; spot flamers can desiccate weeds; anti-transpirants (sold to reduce water use by plants) can also serve as barriers to pathogens such as the rust and mildews that attack plant leaves.

Biological controls. Predators (including lady beetles and lacewings), parasitoids (such as *Encarsia formosa*, a mini-wasp that kills whiteflies) and pathogens (like *Bacillus thuringiensis*, a naturally occurring bacterium that kills caterpillars), all of which are available from nurseries and mail-order suppliers, can suppress many common garden pests.

Least-toxic chemical controls. There are a growing number of pesticides that are only slightly toxic to humans and wildlife, target pests selectively and do little damage to natural enemies. These include insecticidal soaps and horticultural oils, absorptive dusts such as diatomaceous earth and silica aerogel, growth regulators that prevent insects from maturing and reproducing and botanical pesticides derived from plants such as the neem tree, which act as antifeedants and repellents.

An IPM approach allows you to work hand in hand with nature to produce a bountiful garden with room for some of the not-so-beneficial critters that insist on sharing our garden space. IPM is more difficult than just spraying a pesticide at the first sight of a pest. But once you have a bit of experience with IPM methods under your belt, you'll find they become second nature. And as these methods restore your garden's natural balance, the number and severity of pest problems will plummet — leaving you a lot more time to enjoy your garden and reap its harvest.

JUDY WHITE

This damage was caused by a tomato hornworm. You can remove these pests by hand or reduce their numbers with tiny parasitic wasps.

Beverly Drake

The Predator Patrol

Putting Good Bugs to Work in the Garden

BY CASS PETERSON

The most effective pesticides in my garden require no sprayer or other application equipment, are always on hand when I need them and are completely nontoxic when handled properly. They also are easy to store and, best of all, cost nothing.

Granted, it is sometimes a little difficult to distinguish my pesticides from my pests. That's because they are all insects.

In the brave new world of ecologically sensitive gardening, beneficial insects — good bugs that eat bad bugs — are gaining popularity as a way to control destructive insects without chemicals and their often toxic side effects. Unfortunately, too many of these white-hat insects are being marketed just like their chemical counterparts: bottled up and sold as a quick fix. And too many gardeners are buying them, at considerable expense, with unrealistic expectations.

Using insects to combat insects is not a simple matter of exchanging a quart of all-purpose bug killer for a quart of lady beetles. It takes a little advance preparation, a little observation and a bit of patience to get started. The good news is that once the system is in place, it can sustain itself with only minor assistance.

The first thing to remember is that beneficial insects are not rent-a-cops. If you want them to be on continuous patrol in your garden, you will have to provide water, shelter and a continuous supply of food.

CASS PETERSON *is the co-owner and operator of the Flickerville Mountain Farm and Groundhog Ranch, a specialty fruit and vegetable farm, near the village of Dott, Pennsylvania.*

The shy spiders that inhabit every garden are friends, not foes. They capture flea beetles and many other pests in their webs.

Consider the lady beetle, better known as the lady bug, a familiar beneficial insect. Adult lady beetles feed on aphids and some other soft-bodied insects, but the real chowhounds are their grayish-orange larvae. To get the most from your lady beetles, you need to keep them around long enough to lay eggs and produce offspring.

That means having a few aphids around at all times for general snacking.

Here's how it works on our farm: Adjacent to our 12 acres of vegetable beds, we have an alfalfa and clover field. The legumes attract aphids, which in turn attract lady beetles to feed upon them. In late spring, we mow the alfalfa field and the lady beetles migrate into the vegetables. By the time the veggies have been cleared of aphids, the alfalfa and clover have resprouted and begun to attract aphids again. The lady beetles return there and the process is repeated in the autumn.

This system can be duplicated on a backyard scale with a small bed of clover in an out-of-the-way place where it won't have to be mowed too frequently. Because adult lady beetles also enjoy a sip of nectar, we are not too diligent about yanking out the Queen Anne's lace and other nectar-rich weeds. In the backyard, a few dill plants, left to flower, would work as well.

Lady beetles hibernate over winter and require shelter. I usually find them in late winter cozied up in some loose tree bark or nestled under old straw mulch. Most horticulturists counsel strict garden sanitation to avoid overwintering pests, but the fact is that the good guys, too, have to sleep somewhere. If a pile of mulch seems too untidy, at least provide a heap of leaves, which can then be tilled into the garden in the spring.

A season or two of catering to lady beetles should reward you with a good resident population, ready to spring into action as the need arises.

Another commonly known beneficial insect is the praying mantis, which is actually not as useful as most gardeners believe. It is nonselective in its diet and will eat lady beetles and other useful insects as readily as insect pests. But it is a striking insect, and when we find its papery egg cases (they look like miniature hornet's nests) we put them in the raspberries where the adult mantids can feed on Japanese beetles.

Among our most useful beneficials are some that the insectaries do not sell, such as the soldier beetle or Pennsylvania leatherwing. Several years ago, this orange-and-black-striped insect started showing up in large numbers in the flower beds, especially on the statice and baby's breath. Since it seemed to do no harm, I let it be, and discovered only later that it has quite a taste for grasshopper eggs. Our grasshopper troubles are gone.

More recently, we have seen an increase in the numbers of spined soldier bugs, prehistoric-looking creatures with long proboscises. The first I saw had a Colorado potato beetle larva impaled on its snout, and I am told they also like to munch on Mexican bean beetle larvae and cabbage worms.

In the orchard dwells another cabbageworm predator, the white-faced hornet. Providing shelter for hornets is perhaps too much to ask of suburban gardeners, but they do have a voracious appetite for caterpillars, slugs and flies. So do yellow jackets, which is some consolation for the occasional stings we suffer each year.

Much less visible, but no less useful, are the tiny parasitic wasps and shy spiders that inhabit our farm. We rarely see a tomato hornworm that is not carrying on its back a fatal load of white sacs, evidence that the trichogramma wasp is alive and well. Sheetweb spiders, crab spiders and other arachnids capture and feed on leafhoppers, flea beetles and myriad other pests.

The diversity of our farm — something always in flower, something always to eat — helps us maintain a diverse insect population, about which we still know very little. But we no longer approach a strange insect with the automatic notion that it is an enemy to be done away with. Until it proves itself foe, we treat it as friend or at least as a noncombatant.

Diversity is also the key for backyard gardeners. It is unrealistic to expect a flourishing insect population in a

land-
scape that consists of a meticulously groomed lawn, a perfectly weeded flower bed and a carefully tended vegetable garden. Good insect habitat tends to the untidy — a little spot filled with wildflowers, clover and moldering leaves.

Needless to say, a gardener interested in encouraging beneficial insects must use pesticides with a sparing hand if at all. Even the common botanical preparations, such as pyrethrum and rotenone, will kill beneficials as well as destructive insects.

It is useful, too, to have a decent insect identification guide on hand, preferably one with pictures of insects at various stages of their development. Early in my gardening career, I spent most of a year squashing some peculiar-looking bugs before realizing that I was exterminating lady beetle larvae, which look not at all like lady beetles.

BEVERLY DRAKE

The days of this tomato hornworm are numbered. It is covered with the white sacs of the parasitic trichogamma wasp.

WALTER CHANDOHA

The Bambi Factor

BY WALTER CHANDOHA

I live on a small farm in Hunterdon County in rural western New Jersey where commercial and residential development is consuming land at a fast clip. Lots of deer live in the area, too; as we get less and less rural, we get more and more deer. With no natural predators to keep them in check, they proliferate and grow fat on the grain of the few still-active farms, on the shrubs and small trees of newly landscaped homes — and on the vegetables and flowers in my garden.

When we were surrounded by large farms the deer had lots of room to roam and were no big problem. But as their space shrank they became increasingly bold and began to invade my garden and grounds, eating everything in sight. I tried numerous ways to thwart them. Some succeeded, others failed.

When I complained about my plight to an old timer in the area, he suggested a dog to help keep the deer away. So my first line of defense was Kittydog, a big, shaggy Bouvier des Flandre. She was about 75 percent effective. She never harmed the deer; chasing them was play for her. After running them out of the garden, across a field and into the woods, she'd quickly return for lavish praise. After she died, our next dog was Rascal, a small coonhound-type mutt, who was about ten percent effective. He tried to please, but once in the tall grass he couldn't see the deer even when he made his high, arabesque leaps.

Now I was desperate. Out in the country, the county extension service is the best place to go to for help with farm-related problems — and if they can't come up with a solution, they know who can. They told me to call the New Jersey Fish, Game and Wildlife people, who suggested a scent repellant and fencing. The repellant was a dark colored, strong-smelling liquid called Magic Circle. The odor coming from rags soaked with the substance and nailed atop four-foot poles helped keep the deer away. It was about 50 percent effective. To prevent rain from washing the scent away, I covered the rags with inverted flower pots.

Their other suggestions: either a six-

WALTER CHANDOHA, *garden writer and photographer, gardens in Annandale, New Jersey. This article was adapted from a piece that appeared in* Country Living *magazine, July 1991.*

This cute little creature will grow large and fat on the prized trees and shrubs of suburban homes — and the produce of their vegetable plots.

WALTER CHANDOHA

In his never-ending quest to outwit semi-domesticated munchers, the author tried Bye Deer, above, a scent repellant. The flower pot at left covers Magic Circle repellant on the pole. Human hair in panty hose hangs from the shrub.

foot woven wire fence topped with two more wire strands spaced a foot apart for a total of eight feet. Or, an electric fence. A friend of mine finds his electric fence about 90 percent effective. But because I use my garden for many photographic projects, I rejected the idea of either type of fencing.

My next suggestion came from bird photographer Laura Riley: human hair. After collecting it at local barber shops, Ms. Riley stuffs it into sections of old nylon hose, ties the sections into three-inch balls and hangs them near vulnerable plants like her rose bushes. She swears by hair but after trying it in various parts of my garden, I found it ineffective.

Next I tried soap. Saved bits and ends of soap were tied in pieces of nylons and, like the human hair, were hung near vulnerable crops. Didn't work. Next, I drilled holes in brand-new bars of Irish Spring soap (I'd read somewhere that its fragrance was the most effective in repelling deer) and hung them in various parts of the garden. Irish Spring didn't work either.

A hot pepper spray was next. Hot chile peppers and garlic liquefied in a blender and sprayed on the things deer like was

Walter Chandoha

Above: Hinder, a non-toxic spray, proved about 90 percent effective if applied early in the season, before deer discovered the garden's tasty tidbits. The red amaranth at right was devoured in a single night.

supposed to keep them away. No way. If anything, it probably added some zip to the bland vegetables they were eating.

A portable radio tuned to an all-night talk station was also a failure.

I have always questioned the oft-repeated companion planting theory. Grow plant A next to plant B and bug E will stay away from A. I frankly think the theory is not valid but since I like to mix up my garden anyway, I thought I'd give it a try as a deer deterrent. That summer I had lots of alliums, herbs and marigolds interplanted with vegetables all over the garden. The deer ate none of these but they feasted on the peppers, beets, tomatoes, chard, carrots, lettuce, cukes and beans.

Then I thought I had the deer problem solved. They had eaten half a row of beets, but those not touched were adjacent to a row of chrysanthemums which were also uneaten and still being pinch-pruned to make them bushy. I assumed the deer did not bother the far end of the beet row because the scent of the nearby mums repelled them, so I scattered pinched mum foliage over the remaining beets. That night they not only ate the rest of the beets, but the mum plants as well. So much

WALTER CHANDOHA

Deer found the yews above extremely yummy. With three scenic strands of electrified wire on top, the picket fence at left will be about 90 percent effective at deterring deer.

for companion planting as a deer repellant.

Finally, a few years ago, I solved the deer problem with fencing, but used in an unconventional way. At the local farm store I found sturdy wire fencing with a two by four-inch mesh in five- and six-foot heights in 100-foot rolls. Six-foot lengths of the fencing cut off the roll and arched over my wide, raised rows adequately protected any vulnerable crops underneath. The deer could not poke their noses between the fine mesh, and chicken wire placed across the ends kept them from reaching under the arches.

That same fall I found still another 100 percent effective way to keep deer from eating vegetables and flowers: I covered the vulnerable stuff with Reemay or Agronet. Both are feather-light polypropylene sheets used to protect crops from frost and cold down to temperatures in the high 20's. They gently rest atop the plants, admitting light and rain while keeping out cold — and pests like bugs and deer. Lettuce covered with the plastic was untouched by deer but an adjacent row of chard was eaten down to the ground — I had forgotten to cover them with either the Reemay or the arches of wire fencing.

Walter Chandoha

A 100-percent-effective way to keep precious produce safe from deer: lightweight plastic row covers, above, and half circles of wire fencing, right.

Despite my failure with soap and human hair, I'm having some success with commercial scent repellants. Hinder, a soapy spray with an ammonialike smell, is about 90 percent effective if applied early in the season, before the deer have discovered the garden's tasty tidbits. The manufacturer advises reapplication after heavy rains. They also indicate it is non-toxic and can be safely used on food plants.

Another scent repellant called Bye Deer has a sweet smell and comes packed in small green cloth bags about as big as a sack of Bull Durham tobacco. To be effective, the manufacturer says, the bags must be placed in close proximity to the deer food; that is, right where the deer might nibble, not a foot away. So far my tests with Bye Deer have been promising, but I'm still testing.

What with the arches of wire fencing, the plastic row covers and now the scent repellants, I have finally been able to thwart the noshing deer, and the garden is burgeoning. So after all these years of battling Bambi and his family I can once again look kindly at them and welcome them to my farm — as long as they're just sightseeing and will let my flowers and vegetables grow.

ROBERT KOURIK

Drip Irrigation

BY ROBERT KOURIK

Drip irrigation is the most water-conserving way to irrigate just about any plant. Water savings, compared with sprinkler and furrow irrigation, range from 30 to 90 percent, with an average conservation of 50 to 70 percent. Drip irrigation is not just for drought-stricken Western gardens; it produces the most prolific growth and abundant yields in every climate. Gardeners throughout the country who have discovered the secret of drip irrigation are realizing greater harvests, even in areas where summer rain is spotty.

Art Gaus, horticulture specialist with the University of Missouri at Columbia Cooperative Extension, reckons a well-timed drip system could double yields; during the droughts of 1980, '83 and '84, he says, it meant the difference between having a crop or no crop at all. In his own garden, during one season Gaus harvested 32 pounds of

---

ROBERT KOURIK *is the author of* Drip Irrigation for Every Landscape and All Climates *(Metamorphic Press, 1992).*

Left: Drip irrigation produces the most abundant yields in any climate.

bush watermelons from a 4- by 4-foot area with plastic mulch and a drip system, compared to 9 to 16 pounds from the same sized area with conventional irrigation.

Drip irrigation gets its name from the action of emitters — small devices that regulate the flow of water into tiny droplets that slowly water the ground without flooding. Drip emitters form a "wet spot" beneath the soil's surface that differs in shape according to the type of garden soil, ranging from carrotlike in sandy soil to a squat and beetlike in heavy clay soil.

Of all the applications for drip irrigation — from containers to shrubs to trees — the vegetable garden can be the most problematic. For a good drip irrigation design for annual vegetables, the most important criteria are flexibility, thorough water distribution, easy removal for soil cultivation, sturdy hoe- and trowel-resistant tubing and a low visual profile. After 15 years of fiddling with all sorts of drip irrigation tubing and gizmos, I've settled on a streamlined system that meets the criteria and works with row crops, raised beds and boxed beds (see the illustration below).

All drip systems require three essential parts (often called the "main assembly") at the hose-bib, before the actual tubing that emits the slow trickle of water ranging from 1/2 to 2 gallons per hour. First, all hose-bibs should have a metal atmospheric vacuum breaker to prevent siphoning of dirt and manure into the home's water supply. Second, there should be a filter to

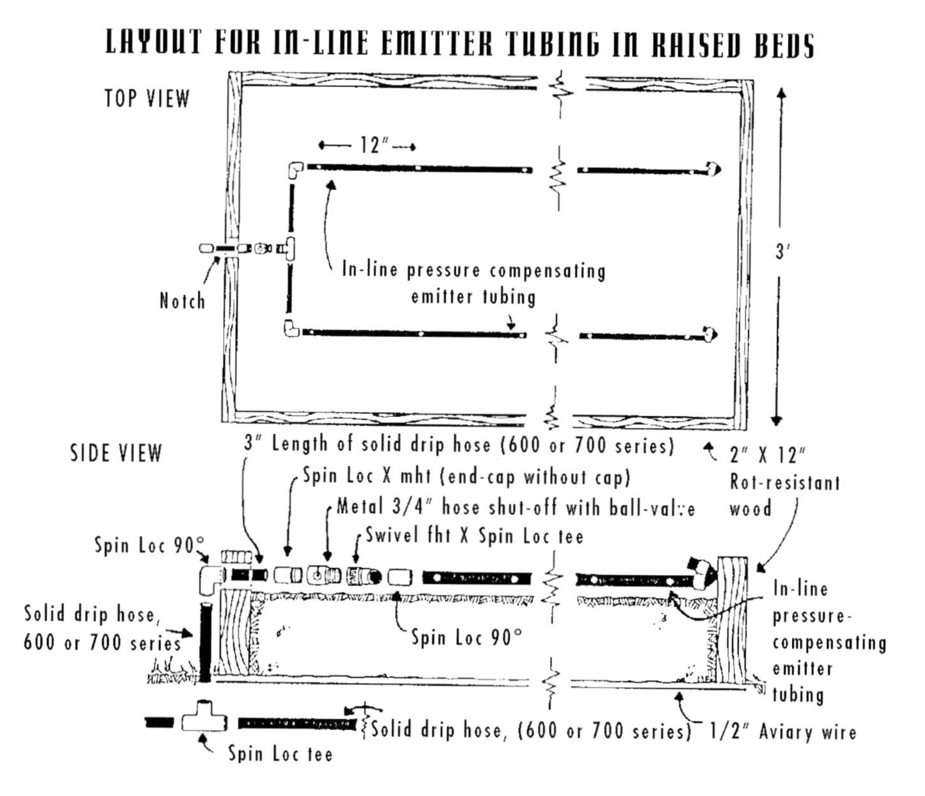

The best part of this system for drip irrigating vegetable beds is the fht (female-hose-thread) swivel tee. With the twist of a wrist, you can separate the in-line emitter tubing from the ball-valve and cultivate without obstruction.

catch any wayward sediment or crud before it can clog the tiny openings of the emitters; a Y-filter, with a ball-valve attached to the end of the chamber for easy flushing, is the best model. And because most municipal and well water supplies are pressurized to 40 to 60 pounds per square inch (psi) — too much pressure for the drip irrigation parts — a pressure regulator, which keeps the pressure at or below 25 psi, should be installed *after* the filter.

Many people think drip irrigation consists of one or two emitters placed near the base of each plant. Such a scheme will keep a plant alive, but will ultimately limit growth because only a small portion of the root zone will receive adequate moisture. The best drip systems have enough emitters so that the wet spots below the soil's surface merge, providing continuous moisture throughout your vegetable plot's entire root zone (The illustration below shows parallel rows of tubing with emitters pre-installed at regular intervals in the black or brown polyethylene hose.) The continuous zone of moisture is 3 to 6 inches beneath the surface, depending upon the soil type. Vegetable transplants can be planted in the dry areas at the surface regardless of the location of the emitters because their roots will be happily exploring the zone of continuous moisture.

Because the emitters release moisture so gradually, the soil's pore spaces don't get waterlogged. The plants' root hairs, not stressed by periodic waterlogging, absorb nutrients more freely, thus improving yields. Michigan State University has documented a 30 percent increase in vegetable crop yields with drip irrigation, even in Michigan's humid climate with abundant summer rain. Drip systems also allow gardeners the option of providing tiny amounts of water on a daily basis to maintain a moist, not wet, soil, which produces the best growth and greatest yields for many crops.

In my garden, I've banned the skinny-diameter "spaghetti tubing" everywhere, with the exception of potted plants. In vegetable plots, spaghetti tubing tangles when you try to remove it for cultivation, is easily damaged by hand and trowel and gets

DRIP IRRIGATION WET SPOTS MERGE BELOW THE SURFACE

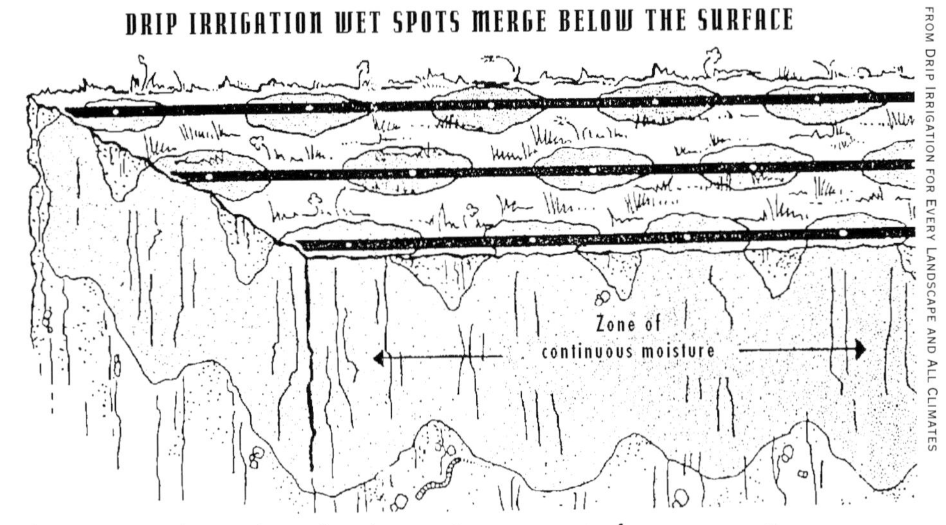

FROM DRIP IRRIGATION FOR EVERY LANDSCAPE AND ALL CLIMATES

The wet spots beneath each in-line emitter merge to form one continuous zone of moisture. The soil for the entire length of the in-line tubing is moist some 4 to 6 inches beneath the surface, depending upon the soil type.

ROBERT KOURIK

pulled out of the 1/2 -inch solid drip hose it's attached to; also, the emitters at the ends of the hydralike tubing can pop off.

My preferred choice of tubing — in-line emitter tubing — has emitters encased within the 1/2-inch tubing at intervals of 12, 18, 24 or 36 inches. These emitters have a mazelike internal path that keeps sediment from clogging them.

The benefits of in-line emitter tubing are many. It is easy to install, suffers less from clogging than porous tubing and some punched-in emitters, works at the greatest range of pressures (10 to 25 psi) and has no external parts to snap off. Also, the connectors, called compression fittings, seal better than the hose-clamps used with porous hose. After almost ten years of use, I have found only a few clogged in-line emitters in hundreds of feet tubing — even with well water high in iron particles.

The drawbacks of the in-line emitter tubing are few: it's not recommended for plants placed far apart or at odd intervals; it can't turn in as sharp a radius as porous tubing can and it doesn't easily germinate broadcast-seeded crops as carrots, mache, beets and turnips. Broadcast-sown crops must be hand-watered until the seedlings have one or two sets of true leaves, then drip irrigation can take over. Large-seeded crops that are planted farther apart, such as beans, squash and sunflowers, can be germinated with the surface wet spots around emitters.

For annual vegetables, the most important aspect of drip irrigation is easy, rapid removal of the tubing to allow for cultivation. This is accomplished by using an "fht" swivel tee, illustrated on page 49. The "fht" stands for "female-hose-thread"; an fht fitting looks just like the end of the hose you connect to a hose-bib. The tee part, resembling the letter T, allows the tubing to make two right-angle turns. This part is connected either to a ball-valve or directly to a garden hose (with a main assembly). To remove the drip system, you simply unthread the fht swivel tee and lift the tubing out of the way. For 3-foot-wide beds with any amount of clay in the soil, only two lines of in-line tubing are required; 4-foot-wide beds might need three lines. If you plan to remove the drip tubing yourself, don't make the beds too long. Lengths of tubing 8 feet-or longer are a bit unwieldy for one person to handle.

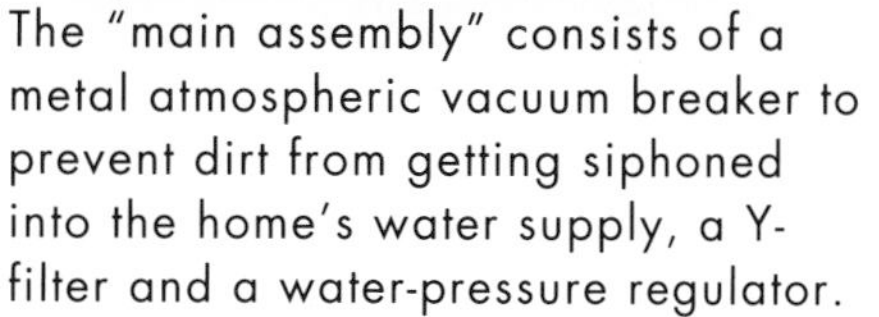

The "main assembly" consists of a metal atmospheric vacuum breaker to prevent dirt from getting siphoned into the home's water supply, a Y-filter and a water-pressure regulator.

A well designed drip system is practically invisible. A thin mulch will hide the tubing before the foliage covers the ground and will protect the plastic from harmful ultraviolet rays.

With your drip irrigation system discreetly in place, you'll be able to fine-tune the amounts of water you apply, whether daily or weekly, to suit the needs of both you and your plants.

Great Greens

BY TERRY KELLER

Greens in the garden offer a variety of tastes and colors throughout the growing season. They are the first vegetables harvested in spring, and can be seeded again in mid to late summer for fall harvests, too. This long season can be extended on both ends of the gardening calendar by taking a cue from the 19th Century French market gardeners and growing greens under glass or, these days, plastic cloches (covers that trap and store radiant energy).

Greens can be loosely divided into two

TERRY KELLER *has been gardening for the past fifty years and is the director of Bronx Green-up, the community gardening program of the New York Botanical Garden.*

PHOTOS BY JOANNE & JERRY PAVIA

categories: the familiar salad greens usually eaten raw in salads (lettuce, arugula, parsley, white mustard and so on) and the potherbs (including collards, beets, spinach and turnip greens), which are usually cooked. Some greens, like spinach, are tasty both raw and cooked.

All greens are easy to grow when the weather is cool (though New Zealand spinach, an Asian vining spinach with a red stem called Basella Malabar and Swiss chard tolerate summer heat), in evenly moist soil rich in organic matter and nitrogen. Most can be seeded directly into the garden as soon as the soil is workable, and germinating plants will withstand lingering frosts in spring and early light frosts in fall.

JERRY PAVIA

Cabbages 'Super Pak', center, and 'Savory King', bottom.

Salad Greens

Salad greens have been cultivated and savored for centuries. One of the oldest greens, arugula (or roquette), was praised by Pliny the Younger thousands of years ago, and was grown by colonists in this country. Today, it's trendy once more.

Arugula should be planted as early as possible in spring, and is best harvested when leaves are four to six inches long. At that point it tastes slightly spicy, like a faintly nutty, mild horseradish. As the weather warms, arugula develops a much stronger flavor — too strong for some. Arugula can be planted again, like many other greens, when the weather cools in fall, and harvest can be extended by lifting plants and moving them indoors to a cool, light spot, or by using cloches. It's easy to make your own simply by covering metal hoops or curved PVC pipe with heavy duty plastic, and they are also readily available through most mail-order seed companies.

Most lettuces also grow best in cool temperatures. Black-seeded Simpson, an early leaf lettuce, emerges quickly when seeded directly into the garden and can be planted every few weeks to ensure a continous harvest before warm weather arrives. However, Buttercrunch, a butterhead variety, Four Seasons, a red-green leaf lettuce, and several head lettuces will tolerate higher temperatures. Seed catalogs usually point out such qualities as likeliness to bolt after the first warm spell.

Lettuces offer many choices and surprises to home gardeners familiar only with grocery store produce. Corn salad (mache), red and green-leafed chicories and radicchio are all lettuces, and all provide different tastes, textures and colors.

Potherbs

Potherbs — the greens to be cooked — are

JOANNE PAVIA

Most lettuces grow best when the weather is cool. Shade helps keep the temperature down.

seeded, grown, cared for and harvested just like the salad greens.

Broccoli raab (or rape) has a sharp but sweet flavor when picked very early in the season. This fast grower should be harvested before its flower buds open into blossoms. Stems, leaves and buds are delicious when steamed with olive oil and slivered garlic. After flowering, broccoli raab should be pulled from the garden and tossed on the garden heap (but do save some of the seeds for next year's crop).

Swiss chard is another easy green to grow. Like most other greens, it tastes best when the leaves are young and tender. Harvest the young outer leaves when they're seven to nine inches long; new leaves will grow out from the center. Swiss Chard tolerates heat well, and a spring planting will produce into early winter.

Kale appreciates cooler weather than Swiss chard; in fact, kale's taste improves with light frosts in fall. Mulch it with hay, or use a cold frame or cloche to insure a crop of kale well into winter. When sowing kale seed, work additional calcium (available from ground limestone) into the soil.

Chinese cabbage and other brassicas are also very cold tolerant. These oriental vegetables will withstand temperatures down into the mid 20s, and with protection will provide greens all winter long. Plant them when the soil cools down at the end of summer.

Many more greens await the adventurous gardener, and most are easy to grow. Cool temperatures, well-prepared and well-drained soil loaded with organic matter and diligent thinning and harvesting will fill your kitchen with greens year-round.

The Upscale Spud

BY ROSALIND CREASY

Purple potatoes for lavender vichyssoise. Rich yellow-fleshed potatoes that appear "prebuttered" when cooked. Classic European potatoes for salade Nicoise. Don't look now, but potatoes have gone "uptown." While the old reliable spuds have long been part of many a comforting meal, specialty potatoes are now adding excitement to garden and table — and a need for a better understanding of our old friend, the potato.

While growing nearly 20 varieties over the last decade, I have found it helps to understand potatoes if you categorize them according to whether they are best baked or boiled. The flaky texture of a good baked potato comes from its "high-solid" content, meaning its large starch molecules and thick skin. Steam forms inside the potato during baking and puffs the starch up; if the steam is allowed to escape, the texture of the cooked potato is light and dry. The large starch molecules also make baking potatoes great for frying and for fluffy mashed potatoes. But when you boil them, the large starch molecules absorb water, expand, soften and break apart — and your potatoes turn to mush.

The best boiling potatoes have small, closely spaced molecules of starch and when boiled, have a waxy, dense texture. Boiling potatoes are superior for stews and potato salad, as they hold their shape well. But when baked, they are fairly heavy and when mashed, sometimes become gummy.

ROSALIND CREASY, *who gardens in northern California, is the author of several volumes, including* The Complete Book of Edible Landscaping *and* Earthly Delights, *both published by Sierra Club Books.*

ROSALIND CREASY

Yellow-fleshed potatoes, right and opposite page, appear pre-buttered when cooked. Purple ones are best steamed to retain their color.

All-purpose potatoes such as Bintje and Yellow Fin can be baked or boiled. When baked, instead of being flaky, they are creamy, and when boiled, they have a soft, starchy quality. The following are some of my favorite varieties.

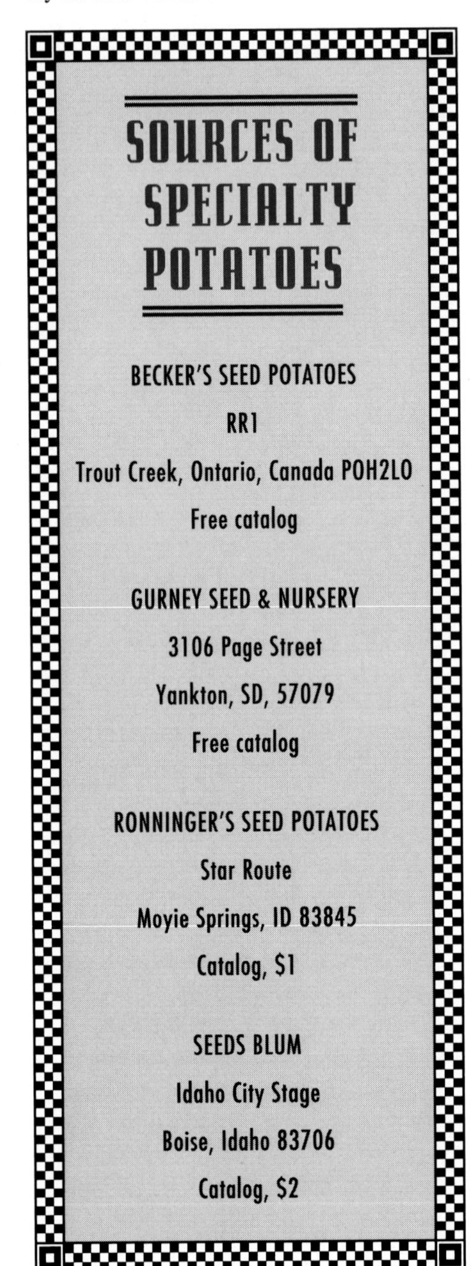

Good Baking Potatoes

Russet Burbank is an oblong, six- to eight-inch potato with brown scaly skin and creamy white flesh. Considered superior, it cooks up dry and flaky. Often used commercially for French fries, Russet Burbank is prone to more diseases than some of the other varieties.

Kennebec, an oblong, four- to five-inch potato, has brown scaly skin and white flesh. When baked, it is flaky and flavorful; it's okay boiled. In the garden Kennebec is disease-resistant and a medium producer, and it stores well.

Good Boiling Potatoes

Pontiac, Red LaSoda and White Rose are readily available. They are of good eating quality and productive in the garden.

Rose Fir/Ruby Crescent is a real treat. It's a fingerling-shaped, pink-skinned potato, four to five inches long and one-and-one-half inches wide, with light-yellow flesh and thin skin. Its flesh is dense, waxy, creamy and almost sweet. Plants produce late in the season and need even watering or tubers become knobby.

All-Purpose Potatoes

Bintje, an oblong, four- to six-inch potato, has cream-colored to yellow flesh. It is good either as a moist, creamy baker or as a medium-textured boiling potato.

Yellow Fin is a two-and-one-half to four-inch long, dumpling-shaped potato with tan skin and creamy yellow flesh that looks buttered inside.

Colorful Potatoes

It is truly startling to open a baked potato and find it bright purple. However, not all blue varieties are equally vivid. Purple Peruvians, for example, are bluer than others. Mineral content of the soil and length of storage also affect color intensity.

All Blue, a four- to six-inch long, oval potato with purple skin, usually has deep-blue flesh, but sometimes the flesh is a disappointing light blue that cooks to dull gray. This is a flaky baker, dry in texture with a mild pleasant flavor. Steamed, All Blue is firm enough for potato salads.

Purple Peruvian, a two- to four-inch long, finger-shaped potato, is usually intensely purple inside and out. Best baked, but passable when steamed, it produces a large sprawling plant with white flowers in the garden.

Steaming is the best cooking method for retaining the color of purple-fleshed potatoes. For boiling, add a few tablespoons of vinegar to the water. When mixing colored potatoes with white ones in a salad, add the colored ones last and mix carefully; otherwise, the colors will look muddy.

ROSALIND CREASY

Plant seed potatoes in spring in a sunny, well-drained spot.

Growing Potatoes

Potatoes grow best in cool weather, so in early spring purchase local standard variety seed potatoes or order specialty varieties from the seed companies listed at left. (Don't plant grocery store potatoes; usually they've been treated with hormones to keep them from sprouting as they often rot before they sprout.)

In spring, select a sunny, well-drained area with good garden soil. Spread three inches of compost and work into the soil. Make two trenches two feet apart, six inches wide, six inches deep and ten to 20 feet long. Cut up potatoes so that each piece contains two "eyes" or indentations. Place pieces in the trench a foot apart and cover with three inches of soil. When the potatoes start to sprout, fill the trench with more soil until it is level with the existing bed. Once the potatoes are growing vigorously, mulch with compost to prevent tubers from turning green. Water regularly and fertilize mid-season with fish emulsion. Avoid high-nitrogen commercial fertilizers as they tend to make the potatoes taste bland and store poorly. Occasionally Colorado potato beetles are bothersome; hand pick or spray them with insecticidal soap.

Storing Potatoes

While some potato varieties store well, others are best eaten soon after harvesting, including Yukon Gold, Bintje and Russet Burbank. Potatoes are best stored in a dark place, at 34 degrees F and high humidity. But they will tolerate temperatures as low as 32 degrees and as high as 50 degrees. The drier the environment, the faster they will shrivel. Remember that if potatoes are stored in any light, the skins will turn green and become toxic if eaten.

The Art & Science of Tomatoes

BY WARREN SCHULTZ

ROSALIND CREASY

When my family raised tomatoes for market in upstate New York, the hardest part was keeping up with the harvest. I should have such a problem now.

Sheer volume was working in our favor then. With 40,000 plants in the ground, some were bound to produce early fruit. And there was no doubt there would be plenty of vine-ripened tomatoes to slice and snack on, and still get enough to market.

But once I left the farm, tomato growing suddenly got more difficult. With room for only a few plants, I learned a tomato-growing truism: the less space you have, the harder it gets.

I learned quickly that our laissez-faire farm technique — planting the same variety year after year, putting them out late enough to avoid frost and letting them sprawl — wasn't the best garden system. So over the years I've experimented with varieties, both heirlooms and hybrids, and techniques from single staking to weaving to caging to sprawling. I eventually realized that there isn't one single best variety (the earliest tomato will never be the biggest or the most flavorful) or one perfect way to grow them.

The answer is to diversify and tailor your variety selection and technique to suit your needs.

You can't contemplate growing tomatoes without considering the question of growing systems. Staking? One stem or two? Caging? Sprawling? Which is the best way to go? The answer is, that depends. Each technique has something to recommend it. Many gardeners have figured out intuitively that pruning and staking provides earlier fruit while untrained plants offer a greater overall yield.

Recent studies at Rutgers University confirm that. Nine plants were grown there in two-foot diameter cages of concrete reinforcing wire. Nine plants were grown with stakes and pruned regularly so that only one or two main stems remained.

WARREN SCHULTZ *is Editor-in-Chief of* National Gardening *magazine and author of* The Chemical-Free Lawn *(Rodale Press, 1989). He grows his tomatoes in northern Vermont.*

And the results? During the first week of picking, the staked tomatoes outscored the caged ones in fruit per plant and yield per plant. Fifty percent more early fruit were picked from the two-stem staked plants than from the caged ones. No surprise there. Because the pruning limits the number of flowers on the plant, there's less competition. The early ripening may also be due to the physical stress of pruning. Stress always makes plants hurry to ripen fruit before they die.

For the same reason, in the final analysis the fruit on the staked plants was larger, by about one ounce per fruit. That's more than a ten percent difference.

In the end, though, the caged plants produced more than twice the tomatoes over the full harvest period. Each caged plant yielded an average of 21 pounds of fruit compared to nine on the single-stemmed plants and eight on the two-stem plants. Because the caged plants are unpruned, they produce much more foliage, which gives rise to more flowers and fruit.

Even before reading these results I had decided to use the two different methods for growing tomatoes.

I plant my early tomatoes in the warmest, sunniest spot in my yard, the bulb and flower bed up against the front of my house. I make sure there's room for

There are scores of tomato varieties, both heirlooms and hybrids. There isn't one best selection — the earliest tomato will never be the biggest or most flavorful. The answer is to try several varieties.

two or three plants between the drying bulb foliage and the emerging perennials. Instead of staking them, I run a pair of strings from the roof to the ground for each plant. As the plants grow, I wrap a stem around each string and pinch out the rest of the suckers.

My main-season crop is planted in cages in the vegetable garden. Besides providing a sturdy structure that can support the most vigorous of plants, cages provide one more advantage. As soon as the plants go in I wrap the cages from top to bottom with clear plastic. Not only does this warm the air inside, but it also cuts down on wind.

Wind causes all kinds of problems for young tomato plants. It bruises the sensitive plants. It breaks the stems. It decreases yield by causing the plant to devote energy into resisting the wind. And it gives rise to diseases: wind carrying particles of sand tears the leaves and makes ports of entry for disease, especially soil-borne bacterial diseases such as *Septoria*. Studies in New Brunswick have shown that adequate wind protection for tomatoes can increase total yield by about 15 percent and early yield by as much as 60 percent.

After preparing the soil with compost or manure, I lay black landscape mat

It's hard to beat the flavor of the heirloom Brandywine tomato.

Husky Gold is a new dwarf tomato suitable for growing in containers.

over the tomato bed. This material has it all over black poly. It's porous, so it allows rain and irrigation water to seep through and reach the soil (and the roots) under it. That's critical for tomatoes, especially if you've ever encountered blossom end rot.

Since I grow the tomatoes about three feet apart I measure that off and cut a hole about two feet in diameter in the mulch for each plant. Early- to mid-May I set the plants out, and drop cages of concrete reinforcing wire over them. Because I first cut the horizontal wires out of the bottom row, the long vertical wires can be stuck through the mulch and into the soil. (I also "weave" a broomstick or other pole through the cage and jam that into the ground.) Finally, I wrap the entire cage, right up to the top, with clear plastic. If frost threatens I can put another piece over top of the cage to protect them further. I leave the plastic on until the air temperature inside consistently cracks 90 degrees.

PHOTOS BY JUDY WHITE

Sweet 100 cherry tomatoes are probably the sweetest of all.

As the plants grow I mulch the bare soil around the stems with grass clippings. Every three or four weeks I lift the mulch and scratch some dried chicken manure into the soil. Occasionally I'll give them a foliar feeding of liquid seaweed or fish emulsion. With cages no pruning or tying is necessary, though if the plants get too lush I sometimes trim out the excess vegetation to let the sunlight in to ripen the fruit.

If frost threatens to shorten the season, as it did in the third week of August, year before last, I just haul out the plastic and rewrap the cages to protect the plants inside. Using this system last year I was eating home-grown tomatoes nearly into November.

I've found this dual system to be just about foolproof — if I've planted the proper variety.

After years of experimenting I've found my early varieties: Early Cascade and Johnny's 361. Others may beat them for sheer speediness, but nothing that I've found can top either (at least in my garden) for their combination of earliness, good fruit size, decent flavor and ability to keep producing all season long. In fact, Johnny's 361 was voted best tasting in a Rutgers taste test.

You might get some argument about that from some people. (No question about it, gardeners love to argue about the best-tasting varieties.) The consensus seems to be that the oldies are the

PHOTOS BY JUDY WHITE

goodies. It's hard to beat the flavor of open-pollinated heirloom varieties, those bred before shipability became the most desirable quality. Probably the one variety touted above all others is Brandywine. Its large fruit have an intense flavor known for its mix of sweetness and tanginess. Other favorites are Bonny Best, Pink Ponderosa and Tappy's Finest.

That's not to say that a variety has to be old and open pollinated to taste good. For flavor in a main-

Besides providing a sturdy structure that can support the most vigorous plants, cages can be wrapped with plastic to protect young tomato plants from damaging winds. Wind protection can increase yields by 15 to 60 percent.

Large areas of tomatoes are often left sprawled. Gardeners with limited

season tomato I still grow some of our farm favorites: the hybrids Moreton and Fantastic. And of course there's Sweet 100, the cherry tomato with probably the sweetest taste of all. But just because the fruits are small, don't think you can get away with growing it in a container. The plants get huge. If you're growing in pots you're better off with a determinate variety such as Pik Red or Toy Boy or the new dwarf indeterminate Husky Gold.

In recent studies, tomatoes were grown with stakes and pruned regularly so that only one or two main stems remained. These plants produced earlier and larger fruits than their caged counterparts.

space should grow their tomatoes with stakes or cages.

Beyond the Green Bell Pepper

BY RENEE SHEPHERD

Until a few years ago, when most Americans thought of peppers, they thought only of those squarish "green peppers" that are put in casseroles, or blanched, stuffed and baked. Or they ate them raw, and found them difficult to digest. Happily, all this is changing, and peppers of all types, sweet as well as hot, are finding a place in the garden. Originating in the New World, both hot and sweet peppers have found their way to almost every ethnic group around the world — from the tapered sweet frying peppers of Italy to the pungent sweet paprikas of Hungary to the tiny and hot spicy chiles of Thailand.

These prolific fruiting plants are lovely and ornamental, with shiny green foliage and pretty white blossoms. Pepper plants do well planted with annual flowers, and are quite handsome when grown in containers. While pepper seeds must be started indoors in most of the climate zones in the United States, their beauty and bounty make them well worth the extra effort.

ROSALIND CREASY

Hot peppers are much more diverse in appearance, flavor and intensity than most Americans realize.

Sweet Peppers

The varieties of sweet peppers available to the home gardener expands every year, spurred initially by the introduction of imported ripe red and yellow Dutch bell peppers for use by American chefs and home cooks who longed for a wider range of fresh vegetables and flavors. We've found out that the standard "green pepper" is actually an *unripe* pepper and will ripen to a deeper color and flavor. Today you can easily obtain seeds for peppers that ripen to gold, yellow, violet, purple, orange and even a rich chocolate color. These ripe colors are matched by the delicious full-bodied sweetness and the thick-walled crunchy flesh that characterizes mature peppers. They also have exceptionally high levels of beta-carotene and vitamin C.

There are several types worth considering for the home vegetable patch. The standard bell varieties come in a myriad of colors. They are usually somewhat blocky, with three or four lobes. All have thick flesh and are eaten raw or cooked. Roasting sweet peppers brings out their sweetness and adds a nutty flavor that is irresistible. Roast them over charcoal or under a broiler. Just brush with good olive oil first

---

RENEE SHEPHERD *is founder and president of Shepherd Garden Seeds, a mail-order seed catalog company specializing in vegetable varieties chosen for the best fresh eating qualities. She travels to Italy and France regularly in search of new varieties, and one of her latest special interests is chiles, both hot and mild, from Mexico, Central America and the Southwest.*

HOW TO GROW PEPPERS

Peppers are a warm-weather crop and they need a long growing season, but starting them indoors will give them ample time to bear fruit in most parts of the country. Sow seed according to directions on the packet, in flats or peat pots filled with sterile soil mix about six to eight weeks before the average last frost date in your area. Peppers of all persuasions need consistently warm temperatures to germinate: about 80 to 85 degrees F is best. Keep seed flats or containers moist but not soggy in a warm place and expect germination in ten to 14 days.

Once seeds have germinated, seedlings immediately need bright light and can take cooler, 70-degree temperatures. Artificial fluorescent shop lights are a good way to provide light for seedlings; suspend them just an inch or two above the tops of plants and move them up as seedlings grow. If using a sunny windowsill, be sure to rotate plants and protect them at night when windowsill temperatures can plummet. Feed seedlings regularly once they get their first true leaves with half strength, all-purpose fertilizer. When plants are about two inches tall, thin or transplant to about three to four inches.

When danger of frost is past, night temperatures are consistently above 55 degrees F and the weather is warm and settled, it's time to transplant your pepper seedlings.

and grill until soft and tender. Then cool and remove the crisp skin.

Lamuyo peppers are more traditional European sweets. They are thick-walled and crunchy with very elongated blunt ends, about twice as long as they are wide. French, Spanish and Dutch gardeners routinely grow these varieties, which can reach six to seven inches in length and are just like American bells.

Eastern European sweet peppers tend to be less blocky and more pimento or heart-shaped, or even round and tomato-shaped. All are used at their ripe lipstick-red stage. The thinner-walled varieties can be dried for grinding up into the sweet full-bodied paprika powders used to flavor meats and vegetable dishes. Thick-walled varieties are often pickled and, as they are not readily available in American stores, it's particularly fun to grow these old favorites.

Bull's Horn frying peppers, also known as Corno di Toro, can grow up to eight inches long with pointed ends, and are traditionally used in Italian, Spanish and Cuban dishes. They have a thinner flesh than bell types and are sliced and quickly sauteed. These peppers are hard to find unless you live in an area with many Cuban or Spanish markets, but well worth cultivating at home.

Hot Peppers or Chiles

Chiles are also members of the *Capsicum* genus. Much more than simply hot peppers, chiles are far more diverse in appearance, flavor and intensity than many Americans realize — until we grow and enjoy them ourselves. Fresh chiles are much tastier than their canned or powdered cousins. They are surprisingly easy to grow in a wide variety of climates and deserve a larger following among gardeners. It's great fun to experiment with several new chile pepper varieties each season, and with the new interest in Asian, Creole, Cajun and Southwest cuisines, cooking with chile peppers has come into its own. Try mild, mellow-tasting chiles as companions to other vegetables, such as green

Plan to set out only the stockiest plants with healthy, well-developed root systems. Pick a sunny spot with well-drained soil that has been amended with ample organic matter. Harden off your seedlings over three to five days to get them used to outdoor conditions. Do this by putting them out in a protected shady spot for first a half day, then a full day, and then move them gradually into full sun. Try to transplant seedlings on an overcast day or in the late afternoon to minimize stress. Space plants 18 inches apart; most peppers or chiles will grow at least several feet tall and need ample room. Plan to stake your peppers, as most modern varieties have a branching habit and heavy fruit sets that will need support. Pepper plants are heavy feeders and should be fertilized frequently — at least once a month with a good all-purpose plant food or a combination of fish emulsion and liquid kelp. They will respond well to a thick layer of mulch applied when they are five or six inches tall. Weed and water regularly and consistently for best crops.

Harvest when fruits are large, glossy and thick-walled, and/or wait for the mature ripe stage when color changes from green and flavor is sweet and full. Ripe colored peppers are also the most nutritious and best for roasting and eating out of hand. When harvesting, always cut, rather than pull, peppers from the plant. Most plants will bear until cool weather takes hold. Harvest regularly for maximum fruit production. Store peppers in the refrigerator in sealed plastic bags but bring to room tempeature before using for best flavor.

Today it's possible to obtain seed for sweet peppers that ripen to gold, violet, orange and other colors.

beans or corn, or with rice or pasta.

Experiment with the flavors of the hotter varieties in pastas, salsas and sauces. When you grow chiles, grow some of their favorite cooking companions — cilantro, corn, beans, garlic, onions, tomatoes, oregano, tomatillos and parsley — so that you have a full palate of flavors to work with in the kitchen. Freshly roasted, skinned and deseeded chiles can be frozen for year-round use, too. (Chiles, like all peppers, do not need blanching before freezing.) When working with chiles, it's a good idea to wear plastic gloves and avoid touching your eyes.

There is much confusion concerning the names of chile varieties. In their native countries, many have one name in their fresh state and another when dried. Sometimes the name depends on how a chile is prepared, or where it is from.

A good basic starting list of available varieties might include the following types to grow from seed:

- Poblano/Ancho, which are aromatic, mild, heart-shaped and deep red/brown. Called Poblano when fresh and Ancho when dried.
- Anaheim or New Mex chiles, also called California chiles. Very mild, six to seven inches long, thick-fleshed, rich and mellow in flavor. Delicious cut into strips and cooked with corn or beans.
- Cayenne, a general name for the thin, hot, dry and pungent pointed little chiles grown for use in French, Creole and Cajun cooking. Coarsely ground Cayenne chiles are also a familiar ingredient in pizza topping.
- Jalapenos, the familiar medium-hot to hot, thick-walled, three-inch-long cylindrical chiles used for a wide range of Mexican and Southwestern dishes.
- Serrano, very hot and spicy two-inch fruits that are very versatile fresh or dried in spicy South American and Asian dishes.
- Habanero, the hottest of the hot, have fiery bright-orange fruits shaped like little tam-o-shanter hats. (Caribbean habaneros are called Scots bonnets.) They are extremely hot and have a fruity quality to their spicy taste.
- Pepperoncini, wrinkly, three- to four-inch light-green or yellow, mildly pungent, thin-walled peppers. Pickle these for delicious, easy-to-make appetizers.

Once you begin to enjoy both the beauty and diversity of both chiles and sweet peppers, you'll want to try many more. There are literally hundreds of cultivars, and growing them opens up a window to the world's cultures and cuisines — a reward available only to the adventuresome gardener.

ROSALIND CREASY

We've found out that the standard green pepper is actually an *unripe* pepper and will ripen on the vine to a deeper color and flavor.

Up With Eggplants

by Alan Gorkin

Tomatoes, tomatoes, tomatoes. Countless articles and books are available on how to grow the perfect tomato, how to stake, when to prune, what kind of heirloom varieties to try and on and on. But what about the lowly eggplant, also a member of the nightshade family, or Solanaceae? With all the reams of material about its cousin, eggplant, more precisely *Solanum melongena esculentum,* has been given a back space in the vegetable garden.

One reason for the situation is a lack of understanding about growing *aubergine,* as eggplant is known in Europe. Eggplants require an even longer season to mature than tomatoes and should be planted as mature plants for best results. Whereas tomatoes can be started five weeks prior to planting out in the garden, eggplant benefits from being started eight to nine weeks prior to planting.

An eggplant in a four-inch pot, well rooted and branched, planted in June in the New York area, for instance, will outperform any planted from cell packs in mid-May. Plants should be kept growing steadily indoors up to planting time with night temperatures above 60 degrees. Avoid crowding or stress from lack of water.

Prior to planting, harden plants off in a cold frame or sheltered spot for several days. When planting, space the plants according to vigor of variety, two to two-and-one-half feet apart in rows two-and-one-half to three feet apart. Subsequent care should consist mainly of weed control and watering. After the ground has warmed and hot weather arrives, this is best done by mulching. If mulch is not used, shallow cultivating will inhibit weed control, but exercise caution as eggplants are surface rooters and deep cultivation can cause harm.

JUDY WHITE

Eggplants now come in a variety of shapes and in white and violet as well as the traditional dark purple.

Watering is best accomplished by deep, periodic soakings to keep the large leaves from wilting.

Harvest the fruits anytime they have attained one third their mature size or until full-sized. To obtain larger fruits, restrict the number of fruits plants can set by removing about 25 percent of the flowers.

In my gardening experience, the main frustration with growing eggplant has been caused by planting too small a plant out too early. Eggplant and peppers (another solanaceous relative) will fare best when planted later as large plants.

ALAN GORKIN, *the former greenhouse supervisor at Old Westbury Gardens in Old Westbury, Long Island, is now the co-owner of Earth Garden, a florist, nursery and garden center in New Canaan, Connecticut.*

Eggplants require a longer growing season than even tomatoes do. For best results, plant mature plants, not cell packs, and plant them a bit late.

The second most annoying problem is insect pests. Flea beetles, small hard-shelled black beetles, can cause serious damage to the tender leaves of young eggplants.

Mexican beetle larvae and Colorado potato beetles are also quite damaging. Row covers, sections of finely woven fabric often used to extend the garden season, can also be loosely draped over newly placed plants to keep out marauding insects. As a bonus, the covers help the tender plants harden off and adjust to increased sunlight. The best preventive measure is to put out large, healthy plants. It never fails: the pests seem to know when I put out weaker young seedlings and set plants back even more.

If further remedies are needed, especially with the tiny and quick flea beetles, you may want to consider applying sabadilla dust according to the directions on the label. Made from the seeds of a tropical shrub in the lily family, sabadilla breaks down quickly, so plants can be treated up to one day prior to harvest. Look for larger insects such as beetle larvae at night with a flashlight and remove them by hand.

Many excellent eggplant varieties are available from seed catalogs. If you are unable to grow your own, buy healthy seedlings from a nursery and pot into four-inch pots. Classic varieties generally found in garden centers include Ichiban, an oriental variety with a lavender to purple, elongated fruit, and Dusky and Black Beauty, with dark purple, glossy, egg-shaped fruits. Grow them for several weeks before planting in the garden and you'll be rewarded with excellent eggplants. Slim Jim is another favorite of mine, but you may have to start this one from seed.

PHOTOS BY JUDY WHITE

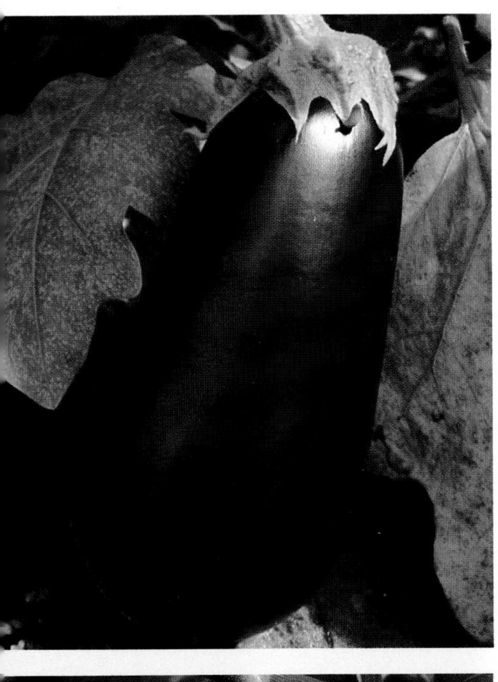

SOURCES

THE COOK'S GARDEN
P.O. Box 535
Londondeery, VT 05148
Catalog $2 (refundable)

JOHNNY'S SELECTED SEEDS
Foss Hill Road
Albion, ME 04910-9731
Catalog $1 (refundable)

SHEPHERD'S GARDEN SEEDS
6116 Highway 9
Felton, CA, 95018
Catalog $2 (refundable)

W. ATLEE BURPEE CO.
300 Park Avenue
Warminster, PA 18974
Free catalog

PARK SEED CO.
Cokesbury Road
Greenwood, SC 29647-0001
Free catalog

ELVIN MCDONALD

JERRY PAVIA

Essential Herbs

BY TOVAH MARTIN

Old-fashioned Sweet Basil

Curley-leaved Parsley

Although herbs may not have quite the nutritional impact of cabbage and squash, although they are not mentioned in the Five Basic Food Groups, they play a crucial role in the culinary world. Where would beans be without basil? Where would tomato paste stand without oregano? A dash of the right herb can coax peas down the gullets of recalcitrant children. Parsley gives a zing to drab baked potatoes; marjoram adds dash to string beans.

In the garden itself, they are wonderful companion plants. Just as the right herb can mellow a rutabaga, many herbs can soften the severe lines of the vegetable patch, forming a bridge between the flower and vegetable domains. Although you wouldn't think of planting a vegetable just because its leaves are blue or its flowers purple, herbs are a different story.

Since herbs already have a reputation based partly on their physical attractions, the temptation to add further visual delights is irresistible. In addition to the standard varieties, hybridizers have engineered variations on the theme: curly parsley, golden sage, purple fennel, creeping oregano, miniature basil and many others. Even if you never eat a single sprig of parsley, you might want to add herbs to the scene simply to relieve the tedium of the vegetable bed.

Although it's been a few centuries since we relied upon herbs for medicine and antiseptics, some are still essential. I can easily live without lovage and I can survive without chervil, but parsley, basil, savory, thyme and oregano are part of my daily diet.

Parsley is a must. I prefer curly parsley rather than the broadleafed Italian type, which tastes too close to celery for my palate. And curly parsley is one of the most ornamental edible herbs in the garden. Lined up in a row (leaving six to eight inches between plants), it makes a fluffy, compact mini-hedge that might even dissuade intruding critters (rabbits love parsley). Swallowtail caterpillars prefer parsley over carrot tops, so planting a lot of it may save your winter crop of root vegetables.

In our Connecticut garden, we start parsley in the greenhouse in February, right along with celery and celeriac. Most herbs germinate in a jiffy, but parsley bears out its bad reputation — that the devil goes to hell and back seven times before this seed will germinate. A toasty environment (65 degrees F) and a lightly moist seed bed will help speed things along, but it's wise to plant seeds indoors if you plan to harvest parsley by summertime. And plant plenty. French, Italian, southern and all sorts of other ethnic recipes call for a handful of parsley to liven up the fare.

Basil is next on the agenda. Most people stick with the plain old *Ocimum basilicum*, the sweet basil you find for sale next to the tomatoes on supermarket seed stands. It's actually the best of the basils, in my opinion. When a recipe calls for basil, chances are that *O. basilicum* is what they have in mind.

And while gourmet seed companies now list half a dozen or more basils with fancy foreign names and tantalizing descriptions, in my experience, most don't live up to their billing: lemon basil has a weak growth habit, lettuce-leaf basil rapidly becomes woody, Fino Verde Compatto

---

TOVAH MARTIN, *is the coauthor of* Keeping Eden *(Bulfinch, 1992) and the author of* The Essence of Paradise *(Little Brown, 1991)*, Moments in the Garden *(William Morrow, 1991) and* Once Upon a Windowsill *(Timber Press, 1988). She is also the staff horticulturist at Logee's Greenhouses, in Danielson, Connecticut, where she and her family grow an acre of vegetables and herbs in their garden.*

tastes like boiled spinach. And who wants to eat camphor basil? The best, in my opinion, are Minimum, Napoletano (alias Mammoth), Green Ruffles and, of course, the no-frills sweet basil.

Although you might add basil frequently to the frying pan, don't go overboard planting basil in the garden. One plump plant, given plenty of room and abundant sun, will easily produce a year's supply of basil if you keep it pinched to promote branching.

On the other hand, you need lots of summer savory. Savory is sort of a bland herb, so you can throw it in casseroles by the handful. But it's a tiny, sprawling affair with rather handsome white flowers scattered among the thin leaves — so the harvest is scant. And because summer savory doesn't rejuvenate after being plucked, we usually plant several crops to keep it coming for meals. Winter savory has earned fame as a perennial counterpart, but though it's just as handsome, it isn't half as tasty.

When there's fresh thyme on hand, it seems as though everything we eat is seasoned with a generous helping of it. English broadleaved thyme, *Thymus vulgaris*, is my favorite for cooking. The leaves are large enough to make a difference tastewise, but they're small enough to be thrown into the pot without dicing. You can also harvest thyme to your heart's content

ELVIN McDONALD

Dwarf oregano, left, and golden oregano, right. Oregano and marjoram are so closely akin that they've become almost synonymous.

TOVAH MARTIN

Summer savory is a subtle herb and doesn't rejuvenate after being plucked, so plant several crops to keep it coming all season long.

and never make a dent in the plant's bushy mass of little leaves. In addition to being user-friendly, *Thymus vulgaris* is perennial. For the best harvest, cut the old branches back to the base in spring, giving the young growth plenty of room to sprout anew.

Oregano and marjoram are so closely akin (both are members of the genus *Origanum*) that, to tell the truth, they've become almost synonymous. *Origanum vulgare* is called both marjoram and oregano in mail-order catalogs and at nurseries. Technically it is wild marjoram and not the best for cooking. *Origanum onites* and *O. majorana*, the gourmet's choices, also share the common name of marjoram, much to the confusion of everyone. The true culinary oregano is *Origanum heracleoticum*, with that inimitable pungent flavor. To get around the confusion with common names, I suggest that you grow several varieties and cook with them according to your taste. They all require sun and they all make neat little mounds in the garden. *O. majorana* is a rather short-lived annual; the rest should survive the winter outdoors, although they can be tilled up with the rest of the garden, replanted and still yield an impressive harvest. All origanums are pungent herbs: a little pinch is enough to add zing to pasta. So don't put in half an acre of either marjoram or oregano but do tuck in a few plants close to the beans.

PHOTOS BY CATHY WILKINSON BARASH, EXCEPT WHERE NOTED

Edible Flowers

BY CATHY WILKINSON BARASH

For years most gardeners were separatists. They fenced in the homely vegetables in the backyard and gave free rein to ornamentals in the front. Commingling of flowers with vegetables or fruits was strictly taboo. As a young child, I railed against these constraints and was always sneaking in some single French marigolds or calendulas with the tomatoes. To my mind, the yellow flowers not only contrasted nicely with the red tomatoes, they also matched the color of the tomato flowers.

Today, of course, many people mix flowers with vegetables, especially those that are edible. You may be growing some already without even knowing it. Pea flowers (not sweet pea, a poisonous ornamental) have a bright, pealike flavor. The flowers are attractive white or purple. Scarlet runner bean flowers are bright orange, contrasting nicely with their cousins, white Dutch runner beans, which have white flowers. Their flavor, unsurprisingly, is reminiscent of the beans. Squash blossoms, used traditionally in Italian cuisine, have also long been a delicacy of Native Americans. Their mild flavor is the perfect foil for stuffing and frying using cheeses, herbs or even meat.

And consider the lowly dandelion, condemned by most as the scourge of the perfect lawn. Grown organically (as all edible flowers must be), the flowers can not only be made into a delectable wine, but their sweetness is a nice addition to an omelet. Once again, the Native Americans appreciated these flowers, making fritters from them.

Edible flowers add color both to the vegetable garden, which can be boringly green, and to the food with which they are served. Some have bright, vibrant colors and flavors to match; others are more delicate, both in color, flavor and size. Most of the culinary herbs have flowers that are edible. And an added bonus for many of the herbs is their ability to ward off winged pests. So consider some of the following — to brighten your vegetable garden as well as your culinary palate.

CATHY WILKINSON BARASH *is a garden writer, photographer and gourmet cook. She has been gardening and cooking since childhood and is the author of* Roses *(Chartwell, 1991) and* Edible Flowers from Garden to Palate *(Starwood Publishing, 1993).*

Jerry Pavia

Calendula & Daylily

Chives

Carnation & Rose

VARIETIES

Common Name	Botanical Name	Color	Flavor
Anise hyssop	*Agastache foeniculum*	Mauve	Licorice
Artichoke (eaten immature)	*Cynara scolymus*	Green	Artichoke, floral
Arugula (rocket)	*Eruca vesicaria*	Off-white	Spicy, peppery
Basil	*Ocimum basilicum*	White, mauve	Spicy, basil
Bean, scarlet runner	*Phaseolus coccineus*	Red-orange	Beany, floral
Bean, Dutch white runner	*P. coccineus* var. *albus*	White	Beany, floral
Bee balm	*Monarda didyma*	Red	Sweet, hot minty
Borage	*Borago officinalis*	Blue	Sweet, cucumber
Broccoli	*Brassica* spp.	Yellow	Mustard, spicy
Calendula (pot marigold)	*Calendula officinalis*	Yellow, orange	(used for color) Mild vegetal (imparted in cooking)
Chives	*Allium schoenoprasum*	Mauve	Chive

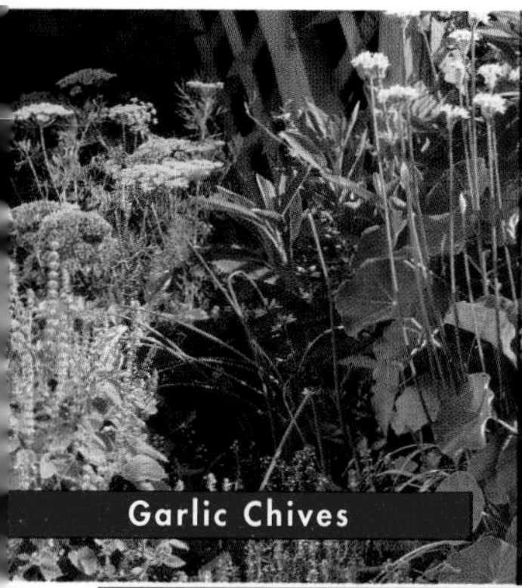

Garlic Chives

Chard, Dill & Johnny jump-up

Lavender

COMMON NAME	BOTANICAL NAME	COLOR	FLAVOR
Chrysanthemum	*Chrysanthemum* x *morifolium*	All colors	Variable, bitter
	'Shungiku'	Yellow, white	Mild
Clove pink	*Dianthus caryophyllus*	Pink	Clove, floral
Coriander	*Coriandrum sativum*	White	Herbal
Daylily	*Hemerocallis* spp.	Yellow, red, orange	Mild, vegetal, sweet
English daisy	*Bellis perennis*	White, pink	Mild vegetal
Garlic chives	*Allium tuberosum*	White	Garlic
Honeysuckle	*Lonicera japonica*	White, yellow	Floral, sweet
Hyssop	*Hyssopus officinalis*	Blue, pink	Slightly medicinal
Johnny jump-up	*Viola tricolor*	Purple, white, yellow	Mild, peppermint
Lavender	*Lavandula* spp.	Purple	Herbal, perfumed
Marigold	(especially *Tagetes tenuifolia* 'Lemon Gem', 'Tangerine Gem')	Yellow, orange	Floral, tarragon

Nasturtiums

Viola 'Cuty'

Chervil, Geranium & Roses

Common Name	Botanical Name	Color	Flavor
Marjoram	*Origanum majorana, O. vulgare*	Pink, lavendar	Herbal, strong
Mint	*Mentha* spp.	White, lavender	Minty, sweet
Mustard	*Brassica* spp.	Yellow	Hot, mustard
Nasturtium	*Tropaeolum majus*	Yellow, red, orange	Peppery
Nodding onion	*Allium cernuum*	Pink	Spicy, oniony
Oregano	*Origanum vulgare*	White, pink	Herbal, savory
Pansy	*Viola* x *wittrockiana*	All colors & mixed	Mild to wintergreen
Pea (garden)	*Pisum sativum*	White, lavender	Pealike, floral
Pineapple guava	*Feijoa sellowiana*	Cream, fuchsia	Sweet, floral
Pineapple sage	*Salvia elegans*	Red	Sweet, herbal
Pinks	*Dianthus* spp.	White, pink, red	Mild vegetal to sweet spice
Radish	*Brassica* spp.	Off white, pink	Mustard, spicy

Tulips

Squash Blossom

Carnation & Thyme

COMMON NAME	BOTANICAL NAME	COLOR	FLAVOR
Red clover	*Trifolium pratense*	Pink	Sweet-strong
Rose	*Rosa* spp.	All but blue	Mild floral, varies
Rosemary	*Rosmarinus officinalis*	Pale blue	Herbal
Sage	*Salvia officinalis*	Blue, purple	Herbal, varies
Savory, winter	*Satureja montana*	White	Herbal, spicy
Scented geranium	*Pelargonium* spp.	Pink, red, white	Variable
Society garlic	*Tulbaghia violacea*	Pink	Garlic
Squash blossoms	*Cucurbita pepo*	Yellow	Mild, vegetal
Thyme	*Thymus* spp.	White, pink	Herbal, varies
Tulip	*Tulipa* spp.	All colors	Mild floral, bean or pealike
Violet	*Viola odorata*	Violet, white	Mild perfume
Yucca	*Yucca* spp.	White	Sweet, mild vegetal

A Cornucopia of Corns

BY FELDER RUSHING

FELDER RUSHING

My dad once stumped some Northern city boys about where the seed for those tiny pickled ears of corn found at salad bars came from. To this day, some of them still think that's what grits are.

What Americans call corn (*Zea mays*), internationally known as maize, is much more than a block of vertical plants in the garden. Being the most efficient of the major grains in transforming the sun's energy into food has made corn one of the four most important food plants on Earth.

About one-half of the world's sweeteners come from corn, as do literally thousands of other products including automobile fuel, cooking oil, beer, corn chips and taco shells, cornbread, millions of cans of whole kernel or creamed corn and, of course, Orville Redenbacher's popped stranglehold on moviegoers. Even a disfiguring corn ear disease called smut (a truly ugly fungal growth) can be eaten as a delicacy.

The importance of corn to the world and its fascinating history are as interesting as growing it is easy: all it takes is sun, seed, fertilizer and water, in about that order. But those top varieties featured in your favorite mail-order catalog or garden center are the result of thousands of years of slow domestication and crude varietal selection. Cultivated as early as 3500 B.C., perhaps the earliest corn was Teosinte, a wild grass with small tassels and diminutive spikelets of seeds.

Corn was one of Christopher Columbus's most important "discoveries." After its introduction to Spain, corn was found cultivated in Africa by 1550. Because of its

JERRY PAVIA

FELDER RUSHING, *a seventh-generation Mississippi gardener, is a garden author, photographer and lecturer. He hosts gardening shows on radio and TV and is a frequent contributor to Brooklyn Botanic Garden's handbooks.*

high caloric value, it helped in enabling European and African populations to swell, and fueling great social changes. The real impact of corn on people, however, was in the livestock arena. Humans ate only the seed of corn; cattle could digest the entire plant. This discovery quickly increased not only the meat and lard supply, but also that of milk, cheese and butter.

Before the early 1800s, people ate mostly young, tender field corn; it has been only in the past couple of hundred years that the sweet corns of today have been widely grown. Corn is now categorized by its color: white, yellow and bicolor (yellow and white kernels on the same ear). Varieties are divided also into those that mature early (in 65 to 70 days), midseason (70 to 80 days) and late (80 days plus). Early varieties are best for northern states; they generally do not make satisfactory growth or ear size in the South.

Super sweet corn is crisp and watery, has four or five times the sugar content of the others and has a slow conversion rate of sugar to starch. Its seeds are small and the plants are slower to establish in the garden than other kinds. However, the ears — which can be eaten raw right in the garden — hold up well on plants and in the refrigerator.

Because the modern marketplace demands a high yield and large, uniform size in corn (or any other commercial vegetable variety), only a few dozen varieties are generally planted each year. The very real risk in this lack of diversity is in susceptibility to disease and other unexpected disasters (prolonged drought, late frosts and so on). For this reason, researchers are trying to identify and maintain for future generations a more diverse gene pool.

From its wild and weedy grass ancestors, we now have 35 or so distinct races of corn, with seed of 11,000 genetically different kinds in storage today in international seed banks.

In addition to the recent interest in such heirloom corn varieties as Santo Domingo Blue (the blue corn of Southwestern cuisine), there are old standbys for every

Corn needs plenty of sun, water and nitrogen fertilizer to grow properly.

FELDER RUSHING

The most efficient of the major grains in transforming the sun's energy into food, corn is one of the four most important food plants on Earth and inspiration for countless cooks — and architects.

region. For your area, any local seed store will carry the most popular kinds (better or worse); the cooperative extension office will have a list of recommended varieties, based on local research. Seed catalogs offer the latest varieties as well, usually with performance comments. Almost all seed sources now carry the normal sweet kinds, sugary enhanced sweets and super sweets.

Over the years, breeders have found that different varieties of corn can tolerate extremes in climate — from the rocky soils of New England, across the muggy Mississippi Delta, to the high, arid southwestern mesas and even to the short, cold seasons of Alaska. Try a new variety or two each year, along with your favorite standby.

Corn seed should be planted after danger of a late frost is past and covered an inch or so with soil. Thin plants to six to 12 inches apart. Plant in blocks of rows for the best pollination. Corn requires more water than most other vegetables: irrigate during dry spells, especially during pollination. Hot, dry conditions during pollination can cause missing kernels, small ears and lower quality.

Corn is a greedy grass which benefits from a regular, even supply of nitrogen fertilizer. Since commercial (chemical) sources of nitrogen generally leach away quickly with irrigation or heavy rains, most gardeners give their corn two side-dressings of nitrogen — one when the plants are a few inches tall and again when knee high. Slow organic fertilizers and compost help smooth out the ride, giving a gentler, steadier feeding than the "feast and famine" of chemicals.

The ancient Mexican method of planting corn, beans and squash together made lots of sense — beans and squash climbed up the cornstalks, while aiding retention of nitrogen. Together they helped keep the soil fertile and arable.

For an ear of corn to develop, powderlike pollen from the tassel at the top of the plant must fall to the sticky silks of the ear, located about halfway up the stalk. This is why corn is best planted closely in several rows, so that pollen won't blow away on windy days. Planting different varieties of corn close together, if they shed pollen at the same time, can cause cross-pollination, resulting in lowered quality or scattered kernels of a different color. Consequently, it is best to separate corn varieties by either distance (several yards, with other vegetables in between) or by maturity time (planting early with late varieties is generally alright).

Pests of corn include earworm, chinch bugs, armyworms, birds and raccoons. Short of using lots of chemicals or following sometimes hit-or-miss natural controls (interplanting with other vegetables and so on), not much can be done to control these creatures. I've heard of farmers-market customers who actually look for a few worms to be sure the corn hasn't been oversprayed with pesticides. Planting lots of corn in several blocks throughout the garden may help reduce overall damage. If a chemical control is needed for earworms (I choose not to use any, and generally get by all right), keep in mind that only the silks need to be treated, and then only for the couple of weeks they are fresh and susceptible to infestation.

Make a note when you first see silks appear from the tips of little ears that the final harvest is generally just under three weeks away. Silks will be withered and dark, ears full and kernels plump and milky-juicy (again, supersweets have clear, watery juice). To harvest, pull down and twist the ears — they'll pop right into your hands. To prevent sugars from rapidly changing to starch, eat or process corn as soon as possible.

By the way, those miniature salad bar cobs are simply immature ears harvested at first signs of silking time, before they can be pollinated. But then what are grits?

The Cultivated Carrot

BY TOVAH MARTIN

Carrots are not top-priority vegetables. When it comes to the hierarchy in the vegetable patch, carrots are ranked right down there alongside potatoes and rutabagas.

But carrots were not always underdogs. In 1603, John Gardiner proclaimed, "if... any city or town be besieged with the Enemy, what better provision for the greatest number of people can be than every garden be sufficiently planted with carrots?" Apparently, the early European settlers in North America were of the same mind, because carrots were among the first vegetables to arrive in this country. Some people claim that Johnny Appleseed carried wild carrot seed with him to scatter on his planting mission throughout the country.

At the time, cultivated carrots were just a step above the wild species from Eurasia. Over the centuries, gardeners selected varieties of *Daucus carota* with a sweet orange taproot while also enhancing its taste and keeping quality. In fact, in the carrot world, the old standbys are still the most popular market varieties. 'Scarlet Nantes', 'Spartan Early' and 'Royal Chantenay', with their long, straight roots and good storage qualities, continue to lead the pack just as they did decades ago. 'Pioneer' and 'Six Pak' have also been around for years and are still valued for their uniform, tender and sweet roots.

Although we aren't inundated with new cultivars, there are some novelties on the market. 'A-plus' is a recent USDA introduction from Wisconsin boasting reddish flesh and an elevated Vitamin A content. 'Trophy' is a very long, slender hybrid with excellent taste, but it tends to break during digging. Several short, stubby, bite-sized hybrids are now available such as 'Kinkol' and 'Parmexl'. But in my opinion, they just don't give you enough carrot for your labor.

Actually, a carrot's quality has more to do with the weather than the name on its seed packet. Carrots are a cool-weather crop. In fact, germination is best when soil temperatures are between 50 and 80 degrees. No matter how carefully you prepare the soil, sprouting will be sporadic when the weather is either too cold or too hot, and the roots develop poorly when the mercury soars.

Equally important to your carrot crop is good, deep, rich soil. Obviously, a plot that is riddled with rocks will impede the downward thrust of your carrot roots, and debris on the soil surface can prevent the tiny sprouts from ever reaching light. So, the first step toward a good crop of carrots is tilling and raking the bed carefully.

The fashion at the moment is to plant carrots in raised beds, and this method works well if you have a small plot to plant. But if you plan to sustain cities with your crop or just feed a family of four over the winter, rows are much easier to weed, harvest and cultivate.

Serious carrot eaters should space their no-nonsense rows two feet apart, allowing ample room for cultivation with a wheel hoe. Plant your carrots shallow and thick — scratching the earth with a hoe handle, sprinkling the seed in the groove and then covering it up with a quarter inch of fine soil.

Even in the best weather, carrot seed germinates lethargically and unreliably. To compensate, it is wise to oversow. Actually, it's nearly impossible to carefully space those tiny seeds the optimal quarter inch apart unless they have been pelleted (coated) for easy handling.

Early carrots can go into the ground in spring along with the first peas. By mid-June, before the weather becomes unbearably hot, the second crop should be safely in place. With a little luck, that crop will keep your family supplied with carrots throughout the winter.

A drought will coincide with your June planting; you can set your watch by it. Watering the carrot crop is a tricky business by anyone's standards, but carrots must have moisture to germinate. A light, gentle rain is ideal, and that is what usually coincides with the spring crop. But in early summer it's either feast or famine in the water department; a sudden cloudburst can drop buckets of water on your fall planting and wreak havoc with the shallowly sown seeds. More often you'll be doing a rain dance — pray for *gentle* precipitation — but if that doesn't work, get out the sprinkler.

Carrots grow slowly at first, especially if the weather is hot. In a searing, dry season, they tend to make pithy, tasteless, pale roots. Of course, you can't do a thing about the weather, but auxiliary watering will increase the soil's moisture content while decreasing the temperature.

Carrots should be thinned to an inch apart immediately upon germination. It's a tedious and sweaty job, but it works wonders with visiting houseguests who have overstayed their welcome. As the roots begin to mature, gradually thin the carrots again to two inches apart using the culls as fingerling snacks.

Carrots are troubled by few pests. Wire worms occasionally make inroads into their roots and nematodes sometimes cause knotty growths on the flesh. Swallowtail caterpillars occasionally dine on a carrot sprig or two. But they do no serious damage and can easily be tolerated in view of their beauty.

Large critters usually pose much more of a threat to the crop. As we all know, rabbits love carrots (they nibble the tops, by the way; I've yet to see a bunny pull a carrot up by the roots and munch away Bugs Bunny-style). Deer can also be a nuisance. Try putting dishes of moth balls (naphthalene) beside the field — the odor should successfully keep everything and everyone at bay.

Carrots are ready for harvest when the roots are about one to one and one-half inches thick. Generally, 55 to 65 days will elapse between germination and maturity. The spring crop often becomes pithy and occasionally goes to seed if left in the ground through hot weather, but the fall crop can stay in the ground throughout the winter. In fact, carrots can be harvested despite the temperatures outside if a thick layer of hay is spread over the bed. Although the traditional method of harvest is simply to pull the crop up by its frilly tops, a preliminary loosening with a spading fork will ease the toil and minimize root breakage.

Carrots can be canned or frozen, but the easiest way to store a large carrot crop is in a root cellar. We line wooden apple crates with sheets of plastic and alternate layers of carrots with moist long-fibered sphagnum moss. Then the crates go into a cellar that remains barely above freezing. If you don't happen to have a root cellar handy, carrots will last for several months in the refrigerator.

Carrots are one of the finest sources of Vitamin A in the vegetable kingdom, with the exception of steamed turnip greens. Tastewise, turnip greens do not compare, and they are certainly of no use as a winter storage crop. In fact, when your garden is nothing but a mound of snow, carrots are still supplying salad with a little spark of color.

INDEX